Aids to Clinical Chemistry

CW00968547

Aids to Clinical Chemistry

D. J. Reynolds
MRCPath
Consultant Chemical Pathologist,
Stoke Mandeville Hospital, Aylesbury,
Buckinghamshire, UK

H. G. M. Freeman
MD, MRCPath
Consultant Chemical Pathologist,
St Peter's Hospital, Chertsey,
Surrey, UK

CHURCHILL LIVINGSTONE
EDINBURGH LONDON MELBOURNE AND NEW YORK 1986

CHURCHILL LIVINGSTONE
Medical Division of Longman Group UK Limited

Distributed in the United States of America by Churchill Livingstone Inc., 1560 Broadway, New York, N. Y. 10036, and by associated companies, branches and representatives throughout the world.

© Longman Group UK Limited 1986

All rights reserved. No part of this publication may be reproduced, stored in a retrieval system, or transmitted in any form or by any means, electronic, mechanical, photocopying, recording or otherwise, without the prior permission of the publishers (Churchill Livingstone, Robert Stevenson House, 1–3 Baxter's Place, Leith Walk, Edinburgh EH1 3AF).

First published 1986

ISBN 0-443-03145-2

British Library Cataloguing in Publication Data
Reynolds, D. J.
Aids to clinical chemistry.
1. Chemistry, Clinical
I. Title II. Freeman, H. G. M.
616.07'56 RB40

Library of Congress Cataloging in Publication Data
Reynolds, D. J. (David John)
Aids to clinical chemistry.

1. Chemistry, Clinical – Outlines, syllabi, etc.
I. Freeman, Heather. II. Title. [DNLM: 1. Chemistry, Clinical – outlines. QY 18 R459a]
RB40.R46 1986 616.07'5 86-12936

Produced by Longman Singapore Publishers (Pte) Ltd.
Printed in Singapore.

Preface

Clinical chemistry has relevance to almost every branch of medicine today. In writing a concise revision work such as this there are inevitable omissions; the lists given are frequently composed of examples to illustrate principles and they are not necessarily comprehensive. Emphasis has been placed on aspects which students commonly find confusing and the space allocated to a topic does not necessarily reflect its relative importance.

The book is intended as a revision aid for senior medical students and for candidates taking the MRCP or part 1 of MRCPath. More detailed text books are available for those seeking further information or a higher qualification.

Aylesbury and Chertsey, 1986

D. J. R.
H. F.

Contents

Introduction

REFERENCE INTERVALS

Laboratories should be able to quote a 'reference interval' for each test performed. This is the range within which the result for that test would fall for 95% of a normal population. It follows that 5% of normal individuals will have a result outside that reference interval. If a large profile of tests is performed on one individual, on average 1 in 20 test results will be outside its reference interval.

The term 'reference range' is often used to mean the same thing but strictly, it is the *full* range of results found when a normal population is tested in order to derive a reference interval, and so will be a slightly wider range than the reference interval.

The older term 'normal range' implies that the results from all normal individuals would be within that range and because of this confusion it should not be used.

N.B. A reference interval applies to a given population and to a specified analytical method used under specific conditions, and therefore, relates only to the laboratory which derived it.

Reference intervals for healthy humans may be affected by:

1. Age e.g. alkaline phosphatase is higher in growing children
2. Sex e.g. creatine kinase is higher in males than females
3. Time of sampling
   (i) diurnal variation e.g. serum cortisol
   (ii) random (day to day) variations e.g. serum iron
   (iii) seasonal variation e.g. plasma 25-OH vitamin D
4. Posture e.g. serum albumin is higher when erect than when recumbent
5. Food intake
   (i) fed vs fasted e.g. plasma glucose, triglycerides
   (ii) content of diet e.g. plasma urea affected by protein intake
6. Race e.g. immunoglobulins higher in African negroes

When age and/or sex have any significant effect on results, age and sex related reference intervals should be used.

INTERPRETATION OF RESULTS

In order to be confident that a result is 'abnormal' or a change is 'clinically significant' one needs to know:

1. the analytical error of the method used
2. the stability of the analyte in health (i.e. physiological variation)

   N.B. considerable variation in serum concentration in health may be related to time of sampling, meals and posture as mentioned above.

SAMPLING PRECAUTIONS

Urine

The accurate timing of urine collections frequently has more bearing on the result than the accuracy of the subsequent chemical analysis.

Serum

1. Avoid taking a blood sample from or near the site of an intravenous infusion
2. Avoid prolonged stasis (which increases the concentration of protein and protein-bound substances).
3. Avoid haemolysis or leaving the sample unseparated for hours. (N.B. samples that have been left unseparated may be more difficult to detect as there is no colour change.)
   Substances which are present in higher concentration in the cells leak into the plasma:
   e.g. potassium
   phosphate
   acid phosphatase
   aspartate aminotransferase
   lactate dehydrogenase
   Storing unseparated blood samples at 4°C slows erythrocyte metabolism and the ion pumps, accelerating the leak from cells to plasma; room temperature is often preferable.
4. Use the correct sample tube
   e.g. fluoride/oxalate for glucose (fluoride inhibits glycolysis by erythrocytes and so preserves glucose, but affects the integrity of the red cell membrane, so its effect is like haemolysis for other analytes).
5. Do not overfill/underfill tubes or tip contents from one tube to another!
   e.g. sequestrene (EDTA) used for haematology is a sodium or potassium salt which prevents clotting by chelating calcium, so contamination with this anticoagulant results in:
   calcium ↓
   sodium OR potassium ↑
   alk. phos. ↓ (EDTA also chelates magnesium which is necessary for alk. phos. activity)

INTERFERING PROCEDURES OR SUBSTANCES

1. Procedures
   (i) prostatic massage may increase the serum level of acid phosphatase for up to a week
   (ii) intramuscular injections may increase the serum level of creatine kinase
2. Drugs etc.
   (i) effects on patients, e.g.
   — sampling for potassium shortly after administration of potassium-losing diuretics (after rapid clearance and before intracellular/extracellular equilibration)
   — dilutional hyponatraemia resulting from drugs such as carbamazepine (the latter appears to increase antidiuretic hormone release)
   — increased serum total thyroxine caused by high-oestrogen contraceptives (oestrogens increase the concentration of thyroxine binding globulin)
   — many drugs which are highly protein bound lower serum total thyroxine by displacing thyroxine from its binding proteins (normal feedback mechanisms keep the free hormone concentration normal)
   (ii) effects on assays, e.g.
   — paracetamol interferes with some glucose assays
   — metronidazole produces spurious lowering of results in some aspartate aminotransferase methods
   — the radio-contrast medium used for intravenous urograms causes a false positive test for proteinuria by the sulphosalicylic acid test

PATIENT IDENTIFICATION

It is essential that the patient is clearly identified on both sample and request form. Never label sample tubes for more than one patient before taking samples; it is surprisingly easy to make a mistake.

There is no point in having accurate results if they are attributed to the wrong patient.

Electrolytes and water

PHYSIOLOGY

Water

1. Intracellular fluid (ICF) — approx. 28 litres
2. Extracellular fluid (ECF) — approx. 12 litres
   (i) Interstitial fluid
   (ii) Intravascular fluid — approx. 3 litres

Water balance

1. Intake
   (i) Oral intake—controlled by thirst
   (ii) I. V. administration of fluids
2. Output
   (i) Renal—controlled by antidiuretic hormone (ADH)
   (ii) Insensible losses
      — sweating, loss in breath, faeces
      — about 500 ml/day but may be much more in a hot environment or if the patient has a fever or is hyperventilating
   (iii) Abnormal losses—see below

Factors controlling ADH release:

1. Osmoreceptors (see Fig. 1)—in general an increase in serum osmolality stimulates ADH release (unless the increase in serum osmolality is caused by a solute which crosses the osmoreceptor cell membrane e.g. urea)
2. Hypovolaemia—this only becomes a stimulus to ADH release in abnormal situations and seems to be a response to life-threatening hypovolaemia

N.B. some drugs may modify ADH release

Osmolality/osmolarity

Osmolality

This is a colligative property and is proportional to the number of

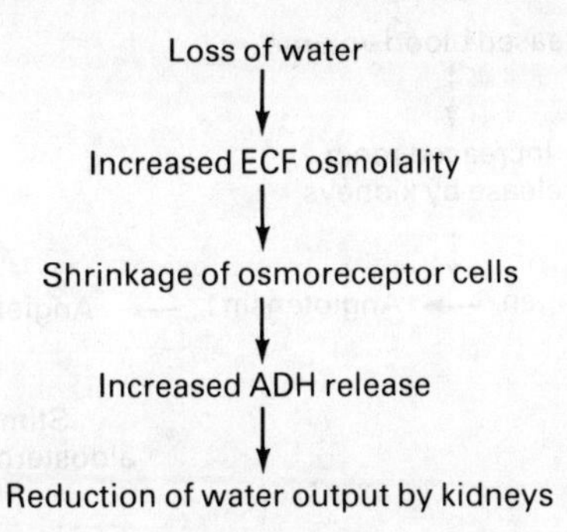

Fig. 1 Normal control of renal water output

particles per *kg of solvent* (i.e. in the case of plasma, per *kg* of *plasma water*)

Units: mmol/kg

Osmolality is directly measured in the laboratory using an osmometer (e.g. by measuring freezing point depression or vapour pressure)

Osmolarity
This is a colligative property and is proportional to the number of particles per *litre of solution* (i.e. in the case of plasma, per *litre* of *whole plasma*)

Units: mmol/l

Osmolarity can be approximated for *plasma* or *serum* (but *not* urine) from the formula:

Plasma osmolarity = 2 × [Na] + [urea] + [glucose]
where all are in mmol/l

In most cases plasma osmolarity calculated in this way is numerically very similar to measured osmolality.

Sodium
1. The majority of sodium is in the ECF
2. Sodium is the major contributor to ECF osmolality
3. ECF osmolality is kept constant by the ADH control mechanism described above such that changes in body water occur in parallel with changes in body sodium

Thus *changes in total body sodium will be reflected by changes in ECF volume* (and hence *blood volume*)

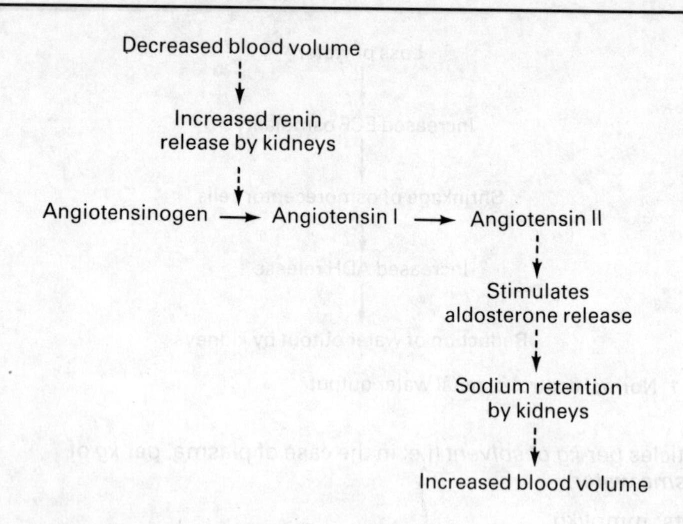

Fig. 2 Renin/angiotensin/aldosterone system

Sodium balance

1. Intake—control of intake is a minor factor
2. Output—under normal circumstances, control of renal sodium output is the major factor in controlling body sodium (and hence ECF and blood volume)

Control of renal sodium output

Three factors are involved:

1. Glomerular filtration rate
2. Renin/angiotensin/aldosterone system (see Fig. 2)
   — affects *distal* tubular sodium reabsorption
3. 'Third factor' (? natriuretic hormone and/or intra-renal haemodynamic changes)
   — affects *proximal* tubular sodium reabsorption

WATER DEPLETION

Pure water depletion is not common—it is more common to find combined sodium and water depletion (see 'Saline Depletion' below).

Water loss will tend to occur initially from the ECF but this will raise the ECF osmolality and so water will immediately be drawn from cells. Thus pure water depletion will involve loss from both ECF and ICF, and therefore, quite a large volume can be lost before there are any significant clinical consequences (namely hypovolaemia).

Causes:
1. Decreased intake of fluid for any reason
2. Increased losses
   (i) from skin
   (ii) from lungs
   (iii) urinary
      a. diabetes insipidus
      b. osmotic diuresis

Severe water depletion gives rise to *hypernatraemia* (see page 11)

WATER OVERLOAD

Severe water overload will give rise to the clinical syndrome of water intoxication.

Features of water intoxication:
1. Somnolence
2. Depression
3. Confusion
4. Anorexia
5. Convulsions } in the most severe cases
6. Coma }

Causes of water overload:
1. Excessive intake—especially iatrogenic (i.v. fluids)
2. Excessive retention
   (i) inappropriate secretion of ADH
   (ii) renal failure

Severe water overload results in *hyponatraemia* (see page 10)

Syndrome of inappropriate ADH secretion (SIADH)
Features:
1. True hyponatraemia (see 'Hyponatraemia' below)
2. Inappropriately concentrated urine
   (urine osmolality > plasma osmolality despite hyponatraemia)
3. Patient in approximate sodium balance
4. No renal or adrenal disease

N.B. though not vital for the diagnosis, a *low serum urate* is a very valuable pointer to SIADH being the cause of an unexplained hyponatraemia.

Causes:
1. Tumours—especially bronchial carcinoma
2. Pulmonary disease—e.g. pneumonia
3. Central nervous system disease—e.g. cerebro-vascular accident
4. ?? 'sick cell syndrome' (see 'Hyponatraemia' below)

SALINE DEPLETION

Pure sodium depletion virtually never occurs—it is almost always accompanied by simultaneous water depletion (hence the use of the name 'saline depletion').

Losses occur from the ECF, and since they are usually isotonic, no fluid is drawn from the ICF. Thus, comparatively small losses lead to hypovolaemia (cf water depletion).

Clinical features of saline depletion:

1. Clinical 'dehydration'
   — this refers to reduced tissue turgor, dry tongue and reduced intra-ocular tension. It is a bad term because it implies water depletion and, as explained above, water depletion will only lead to these features when very severe
2. Tachycardia
3. Hypotension

(↓ increasing severity of saline depletion)

Causes:
Virtually always increased losses

1. G. I. tract
   (i) vomiting
   (ii) diarrhoea
   (iii) fistulae, etc.
2. Urine
   (i) diuretics
   (ii) adrenal hypofunction
3. Sweat

Isotonic losses will not affect the plasma sodium until severe hypovolaemia occurs (hypovolaemia is a secondary stimulus to the release of ADH and thus severe saline depletion may be associated with hyponatraemia).

SALINE OVERLOAD

Saline overload results in expansion of the ECF and hence, leads to hypertension and/or oedema.

Causes:
Usually decreased output

1. Renal failure
2. Oedema states
   (i) congestive cardiac failure
   (ii) cirrhosis
   (iii) nephrotic syndrome

3. Cushing's syndrome
4. Conn's syndrome (primary hyperaldosteronism)

Sodium retention is usually accompanied by simultaneous water retention and so in most cases, plasma sodium remains within the reference interval. However, if water retention exceeds sodium retention (as may occur in renal failure or severe oedema states), a dilutional hyponatraemia may occur.

OEDEMA

Causes of generalised oedema:
1. Increased venous pressure e.g. congestive cardiac failure
2. Low serum albumin
   (i) nephrotic syndrome
   (ii) cirrhosis
   (iii) protein-losing enteropathy

Pathophysiology
1. Altered capillary pressures—an increase in venous hydrostatic pressure (congestive cardiac failure) or a decrease in oncotic pressure (low serum albumin) will alter the balance of water movement out of the capillaries at the arterial ends and movement in at the venous ends (see Fig. 3).
2. Secondary hyperaldosteronism:

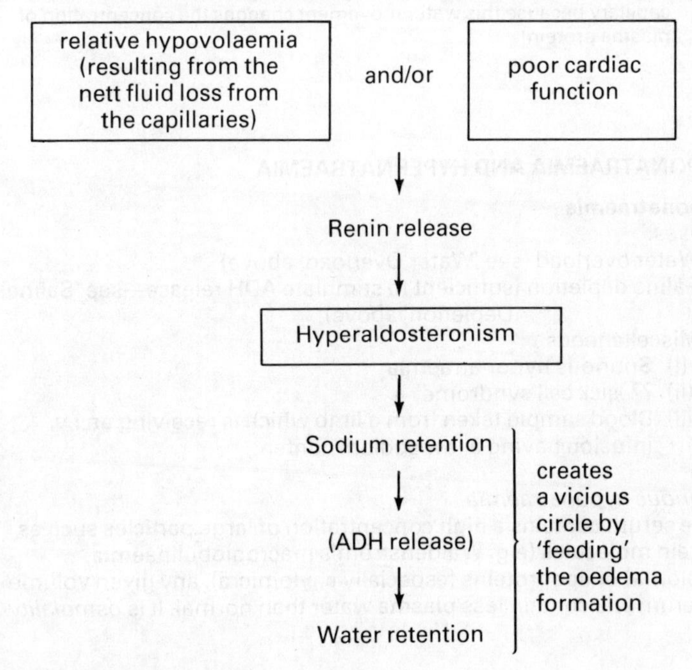

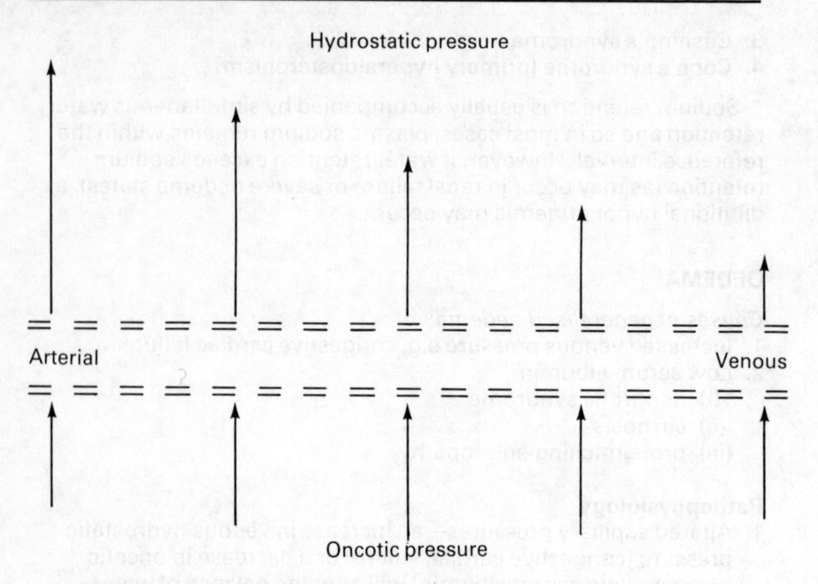

Fig. 3 Starling's Law of the Capillary
Water moves out at the arterial end and back in at the venous end according to the balance between hydrostatic and oncotic pressures. (The oncotic pressure changes slightly along the course of the capillary because this water movement changes the concentration of plasma protein).

HYPONATRAEMIA AND HYPERNATRAEMIA

Hyponatraemia

Causes:

1. Water overload (see 'Water Overload' above)
2. Saline depletion (sufficient to stimulate ADH release—see 'Saline Depletion' above)
3. Miscellaneous
   (i) Spurious hyponatraemia
   (ii) ?? 'sick cell syndrome'
   (iii) Blood sample taken from a limb which is receiving an i.v. infusion having a low sodium content

Spurious hyponatraemia

If the serum contains a high concentration of large particles such as protein molecules (e.g. Waldenström's macroglobulinaemia, myeloma) or lipoproteins (especially chylomicra), any given volume of serum will contain less plasma water than normal. It is *osmolality*

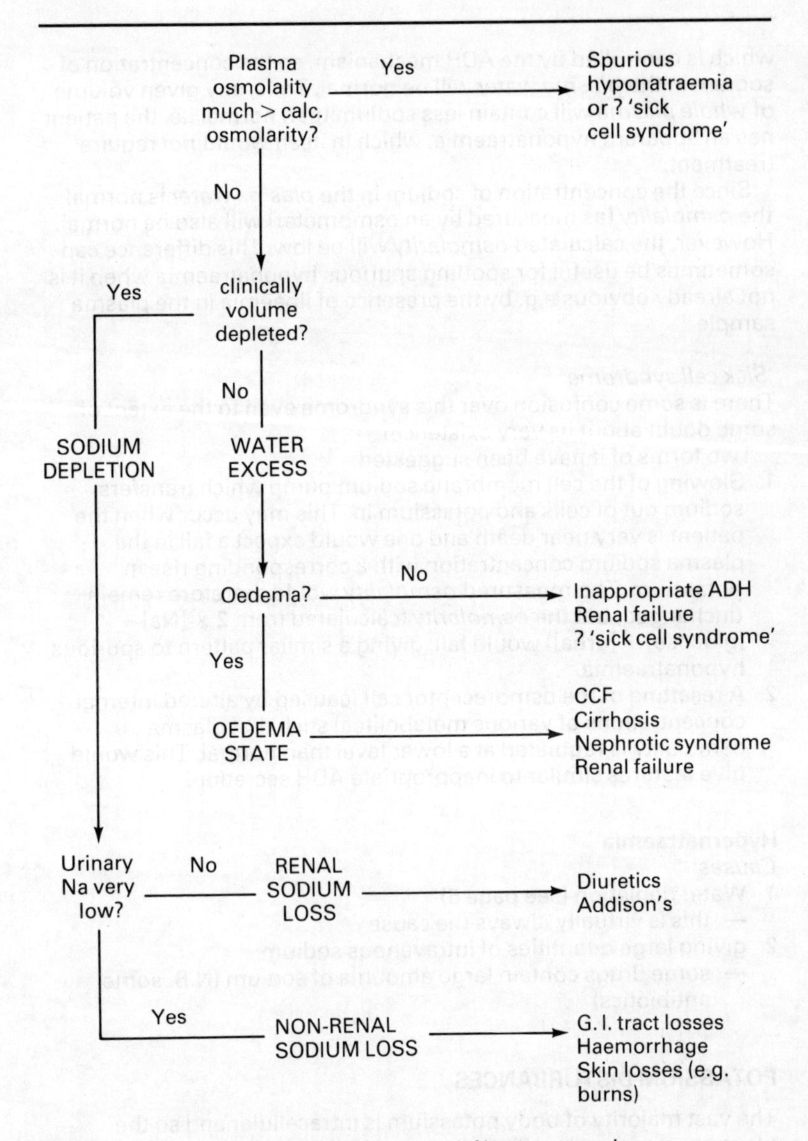

Fig. 4 Flow chart for diagnosis of the cause of hyponatraemia

which is controlled by the ADH mechanism, so the concentration of sodium in the *plasma water* will be normal. Thus, any given volume of *whole plasma* will contain less sodium than normal i.e. the patient has an apparent hyponatraemia, which in itself would *not* require treatment.

Since the concentration of sodium in the *plasma water* is normal, the *osmolality* (as measured by an osmometer) will also be normal. However, the calculated *osmolarity* will be low. This difference can sometimes be useful for spotting spurious hyponatraemia when it is not already obvious e.g. by the presence of lipaemia in the plasma sample.

'Sick cell syndrome'
There is some confusion over this syndrome even to the extent of some doubt about its very existence!

Two forms of it have been suggested:

1. Slowing of the cell membrane sodium pump which transfers sodium out of cells and potassium in. This may occur when the patient is very near death and one would expect a fall in the plasma sodium concentration with a corresponding rise in potassium. The measured *osmolality* would therefore remain unchanged, but the *osmolarity* (calculated from 2 × [Na] + [glucose] + [urea]) would fall, giving a similar pattern to spurious hyponatraemia.
2. A resetting of the osmoreceptor cell (caused by altered internal concentrations of various metabolites) such that plasma osmolality is regulated at a lower level than normal. This would give features similar to inappropriate ADH secretion.

Hypernatraemia
Causes:

1. Water depletion (see page 6)
   — this is virtually always the cause
2. giving large quantities of intravenous sodium
   — some drugs contain large amounts of sodium (N.B. some antibiotics)

POTASSIUM DISTURBANCES

The vast majority of body potassium is intracellular and so the plasma potassium level is only a poor indicator of total body potassium. Indeed, plasma potassium concentration can be positively misleading if there is a disturbance of the normal mechanism controlling the balance between intracellular versus extra-cellular potassium e.g. with acid-base disturbances—thus in diabetic ketoacidosis there is usually a total body potassium deficit but the plasma potassium may be *high* before treatment.

Relationship between acid-base disturbances & potassium
Normally, alkalosis causes hypokalaemia and vice versa. The mechanism for this is two-fold:

1. In the distal renal tubule, sodium is reabsorbed in exchange for either potassium or hydrogen ions. If potassium is deficient more hydrogen ion is lost into the urine and vice versa.
2. The intracellular/extracellular balance of potassium and hydrogen ion is similarly affected.

Acidosis and hyperkalaemia bear a similar relationship to each other.

N.B. The combination of hypokalaemia with acidosis is rather unusual and points to a simultaneous loss of potassium and bicarbonate e.g. a renal tubular disorder like renal tubular acidosis.

Hypokalaemia
Features:

1. Cardiac
   — ECG abnormalities and cardiac arrhythmias
   — potentiates the toxic effects of digoxin
2. Renal
   — prolonged hypokalaemia impairs the ability of the tubule to reabsorb water → polyuria
3. Muscle
   — skeletal muscle weakness
   — impaired recovery from post-operative paralytic ileus

Causes:

1. Decreased intake or absorption—(*rare*)
   (e.g. malabsorption syndrome)
2. Increased losses
   (i) Urinary
      a. Hyperaldosteronism
      b. Diuretics
      c. Alkalosis
      d. Cushing's syndrome
   (ii) G. I. tract
      a. Vomiting
      b. Diarrhoea
      c. Laxative abuse
      d. Naso-gastric tube losses, fistulae etc.
3. Transfer from ECF to ICF
   (i) Treatment of diabetic coma
   (ii) Alkalosis
   (iii) Stress (e.g. surgery, myocardial infarction)
      — almost certainly related to increased catecholamine levels
   (iv) Familial periodic paralysis (*rare*)

Hyperkalaemia
Features:
Cardiac arrest—the higher the plasma potassium, the greater is the risk

Causes:
1. Excessive intake (*rare*)—usually iatrogenic
2. Decreased output
   (i) Renal failure
   (ii) Adrenal insufficiency
3. Release from cells
   (i) Catabolic states
   (ii) Acidosis

MAGNESIUM DISTURBANCES

Distribution of magnesium:

Extracellular fluid	—	1%
Intracellular	—	35%
Bone	—	64%

Magnesium balance:
The main homeostatic mechanism is control of renal excretion but little is known of how this is regulated.

Hypermagnesaemia
The most important cause of hypermagnesaemia is *renal failure.*
Magnesium excess has been blamed for some of the neurological complications of chronic renal failure.

Hypomagnesaemia
Causes:
1. Decreased intake/absorption
   e.g. malabsorption, malnutrition, alcoholism
2. Increased losses—gastrointestinal or urinary
   e.g. vomiting, overuse of diuretics
3. Diversion from plasma to bone
   e.g. after parathyroid surgery (especially in patients with severe hyperparathyroid bone disease)

Features:
1. Neuromuscular dysfunction → hyperexcitable state (fits, tetany, tremor, muscular weakness, tendency to cardiac arrythmias)
2. Hypocalcaemia
   — probably related to impaired parathyroid hormone release
   — unresponsive to treatment unless the magnesium depletion is corrected

POLYURIA

Causes:

1. Diabetes mellitus (osmotic diuresis caused by hyperglycaemia)
2. Chronic renal failure (see page 25)
3. Hypercalcaemia*
4. Hypokalaemia (prolonged)*
5. Primary polydipsia (often psychogenic)
6. Diabetes insipidus (see page 85)
   - (i) Cranial
     — deficiency of ADH
   - (ii) Nephrogenic*
     — renal resistance to the action of ADH
7. Drugs (e.g. lithium)*
8. Fanconi syndrome*

* See 'Renal Tubular Disorders' page 28

Investigation:

1. Confirm polyuria—before admitting a patient confirm that the urine output is in fact $>$ 2.5 litres/24h
2. Measure plasma glucose, urea (and/or creatinine), calcium and potassium levels to exclude causes 1 to 4 above
3. Test for glycosuria and measure serum phosphate and urate to exclude Fanconi syndrome
4. Water deprivation test
   — to differentiate primary polydipsia from diabetes insipidus
5. Desmopressin test
   — to differentiate cranial from nephrogenic diabetes insipidus

(4 and 5: see appendix)

HYPERTENSION

Causes:

1. Essential (idiopathic) hypertension
2. Secondary hypertension
   - (i) Renal disease
   - (ii) Endocrine disease
     - a. Hyperaldosteronism
     - b. Cushing's syndrome
     - c. Phaeochromocytoma
     - d. Adrenogenital syndrome
     - e. Thyrotoxicosis (severe)
     - f. Acromegaly
   - (iii) Pre-eclampsia
   - (iv) Coarctation of the aorta

Hypertension associated with hypokalaemia

Causes:

1. Treatment of the hypertension with diuretics
2. Hyperaldosteronism (see page 82)
   (i) Primary (i.e. Conn's syndrome)
   (ii) Secondary
      a. Renal artery stenosis
      b. Malignant hypertension
3. Cushing's syndrome (see page 80)

Kidney

PHYSIOLOGY

Glomerulus

Glomerular filtrate produced by ultrafiltration:

1. Approx. 150 litres/day
2. Iso-osmolar with plasma
3. Macromolecules not filtered
   — threshold size equivalent to the albumin molecule i.e. molecular weight of 65 000–70 000

Proximal tubule

Active reabsorption of:

1. Glucose
   — normally all reabsorbed unless plasma glucose higher than threshold level which is usually about 10 mmol/l
2. Sodium
   — approx. 80% reabsorbed
   — influenced by 'third factor' mechanisms (see page 6)
3. Potassium
   — approx. 80% reabsorbed
4. Bicarbonate
   — results indirectly from the active secretion of hydrogen ion (see Fig. 5)
5. Phosphate
   — affected by parathyroid hormone
6. Aminoacids

Passive reabsorption of:

1. Water
2. Chloride

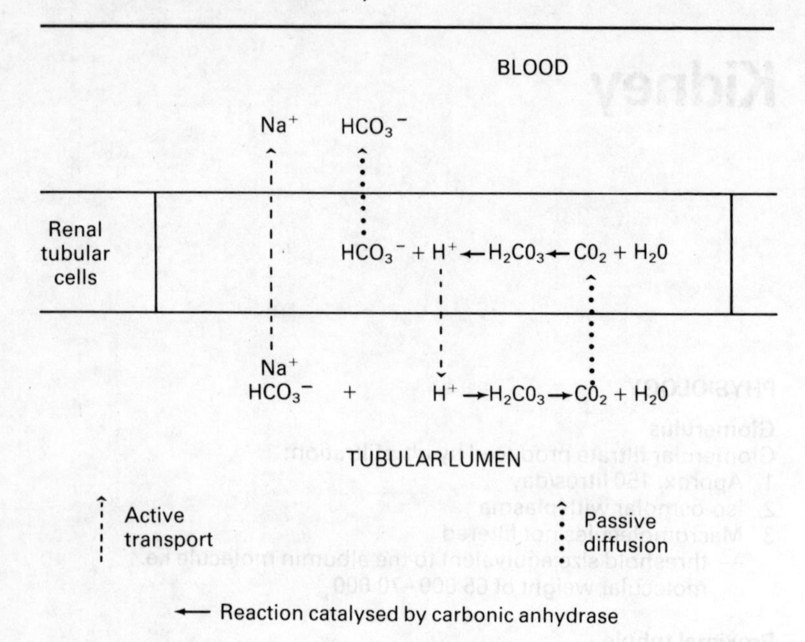

Fig. 5 Renal tubular reabsorption of sodium bicarbonate

Loop of Henle

1. Sodium chloride reabsorption
   — 15% of filtered sodium reabsorbed here
2. Production of hyperosmolar renal interstitium
   — the osmolality increases towards the tip of the loop.
   — uses a counter-current multiplier mechanism.
   — vital for the production of concentrated urine in the collecting ducts

Counter-current multiplier mechanism (see Fig 6)
Though tubular fluid passes through the descending limb of the loop before the ascending limb, the processes occurring in these limbs will be described in reverse order to enhance understanding.

Ascending limb:
— impermeable to water
— active pumping of chloride from the tubule into the interstitium (with sodium following passively) leads to trapping of sodium chloride in the medullary interstitium

Descending limb:
— passive movement of sodium into the tubule and water out under the influence of the increasing sodium chloride concentration in the interstitium on moving towards the tip of the

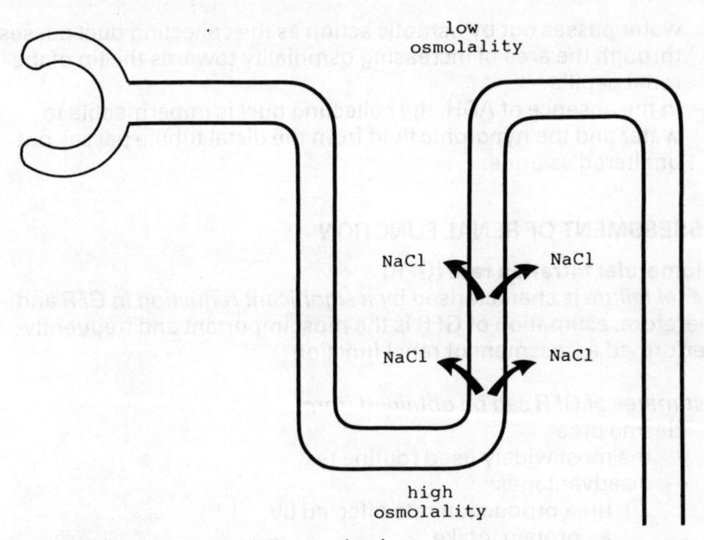

Fig. 6 Counter-current multiplier mechanism.
The important function of the counter-current multiplier mechanism is the production of a hyperosmolar renal interstitium. Sodium chloride is pumped out of the ascending limb of the loop of Henle. Because of the loop structure of both the nephron and the vasa recta, the sodium chloride is not 'washed away' and so becomes 'trapped' in the interstitium, thus producing a high osmolality.

loop (this increasing concentration having been produced by active pumping from the ascending limb)

Vasa recta:
— these blood vessels which supply the medulla are also loops and they follow the course of the loops of Henle. Their looped structure minimises the 'washing-out' of sodium chloride from the interstitium which would otherwise occur.

Distal tubule
1. Tubular fluid entering the distal tubule is hypotonic to plasma because of the active pumping occurring in the ascending limb of the loop of Henle.
2. Sodium reabsorption in exchange for potassium and hydrogen ion
   — fine tuning of sodium excretion under the influence of aldosterone (see page 6)

Collecting duct
Antidiuretic hormone (ADH) acts here:
— it increases the permeability of the collecting duct to water, so

water passes out by osmotic action as the collecting duct passes through the area of increasing osmolality towards the tip of the renal papilla
— in the absence of ADH, the collecting duct is impermeable to water and the hypotonic fluid from the distal tubule passes out unaltered as urine.

ASSESSMENT OF RENAL FUNCTION

Glomerular filtration rate (GFR)

Renal failure is characterised by *a significant reduction in GFR* and therefore, estimation of GFR is the most important and frequently performed assessment of renal function

Estimates of GFR can be obtained from:

1. Plasma urea
   — the most widely used routine test
   — disadvantages:
   (i) urea production rate affected by:
   a. protein intake
   b. cellular catabolism
   (ii) plasma urea is affected by urine flow rate (because urea is reabsorbed by the tubule and so more is reabsorbed at low flow rates)
2. Plasma creatinine
   — creatinine production rate is proportional to skeletal muscle mass
   — if muscle mass is constant, serial measurements provide very good assessment of renal function
3. Clearance measurement (see Fig. 7)
   (i) Creatinine clearance
   This is an example of a 'steady-state' clearance. Such tests depend on the plasma concentration of the compound being measured remaining constant during the test, and in the case of creatinine this is normally true because it is produced at a constant rate by muscle. A timed urine collection (normally 24 hours) and a simultaneous blood sample must be obtained.
   N.B. Creatinine clearance is not ideal for GFR determinations because creatinine is secreted by the tubule to a small extent, thus it tends to over-estimate GFR especially at low GFR's.
   (ii) Chromium-51-labelled EDTA clearance
   Normally performed as a 'single-shot' technique in which a bolus of the compound is injected intravenously and the plasma decay curve is plotted by counting the radioactivity in serial blood samples. From this the GFR can be calculated mathematically.

For the clearance of a compound to reflect the GFR it should be *freely filtered* by the glomerulus and be *neither secreted nor reabsorbed by the tubule.*

If it is freely filtered:

Rate of filtration of compound = GFR × concentration in plasma

If it is neither secreted nor reabsorbed by the tubule:

$$\text{Rate of filtration of compound} = \text{Rate of urinary excretion of the compound}$$

$$= \frac{\text{concentration}}{\text{in urine}} \times \frac{\text{urine flow}}{\text{rate}}$$

Therefore:

$$\text{GFR} = \frac{\dfrac{\text{concentration}}{\text{in urine}} \times \dfrac{\text{urine flow}}{\text{rate}}}{\text{concentration in plasma}}$$

Fig. 7 Calculation of GFR from clearance measurements

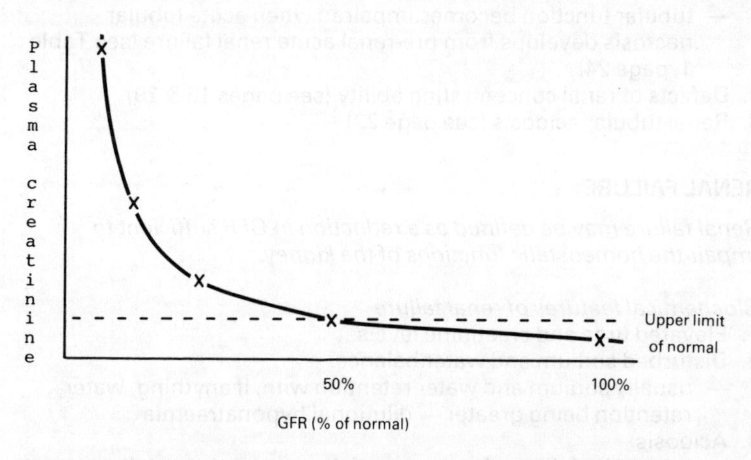

Fig.8 Relationships between plasma creatinine and GFR. *N.B.* Plasma creatinine doubles as the creatinine clearance halves

Creatinine clearance versus plasma creatinine

It will be seen from Fig. 8 that plasma creatinine may be normal until creatinine clearance has dropped to 50% of normal (i.e. for a patient whose plasma creatinine concentration started at the low end of normal). Also, the curve is fairly flat until creatinine clearance has fallen below 50% of normal. Thus it is in cases where the plasma creatinine is normal or slightly elevated that creatinine clearance is worthwhile. After that, it is preferable to monitor renal function by means of serial plasma creatinine measurements (this also overcomes the inaccuracies of 24 hour urine collections).

The glomerular filter
Leakiness of the glomerular filter is detected by:
1. Proteinuria—its presence and quality
2. Haematuria
3. Examination of the urinary deposit

Selectivity of proteinuria
— the greater the damage to the glomerular filter, the higher will be the molecular weight of proteins 'leaked' into the urine
— this can be assessed by comparing the clearance of two proteins of different size, often transferrin and IgG.
— the ratio of these clearances is expressed as a *selectivity index*

Tubular function
Tests of tubular function are not performed very frequently but are of value in:
1. Acute renal failure
   — tubular function becomes impaired when acute tubular necrosis develops from pre-renal acute renal failure (see Table 1, page 24)
2. Defects of renal concentrating ability (see pages 15 & 29)
3. Renal tubular acidosis (see page 29)

RENAL FAILURE

Renal failure may be defined as a reduction in GFR sufficient to impair the homeostatic functions of the kidney.

Biochemical features of renal failure
1. Elevated urea and creatinine levels
2. Disturbed sodium and water balance
   — usually sodium and water retention with, if anything, water retention being greater → dilutional hyponatraemia
3. Acidosis
   — caused by failure of excretion of the acid load normally excreted daily by the kidneys
4. Potassium disturbances
   — usually potassium retention → hyperkalaemia
5. Calcium and phosphate disturbances
   — usually phosphate retention
   — often hypocalcaemia caused by
      (i) acidosis reducing the level of protein-bound calcium in serum

(ii) high phosphate level within renal tubular cells inhibits production of 1,25-dihydroxycholecalciferol → poor calcium absorption from the gut

— hypocalcaemia → secondary hyperparathyroidism
— in chronic renal failure, if these changes are not prevented, they lead to a metabolic bone disease called *renal osteodystrophy*

6. 'Middle molecule' retention
— compounds with molecular weights of 1000–2000
— not routinely measured
— believed to be responsible for many of the symptoms of 'uraemia' (malaise, nausea etc.)
7. Urate retention
8. Secondary hyperlipidaemia (see page 153)
9. Magnesium retention → hypermagnesaemia

Acute renal failure (ARF)

Causes

1. Pre-renal
— poor renal perfusion because of hypovolaemia and/or reduced cardiac output
— recovers if renal perfusion restored
2. Renal
(i) 'Acute tubular necrosis'
— occurs if pre-renal failure is not corrected fairly quickly
— a better name would be 'acute ischaemic renal failure'
— see below for differentiation of this from pre-renal failure
(ii) Glomerulonephritis
— especially rapidly progressive (i.e. crescentic) glomerulonephritis (e.g. Goodpasture's syndrome)
(iii) Nephrotoxins and drug reactions
(iv) Disorders of the renal vasculature
a. large vessels
— renal artery (embolism, thrombosis etc.)
— renal vein thrombosis
b. small vessels
— disseminated intravascular coagulation
— thrombotic microangiopathy (e.g. haemolytic uraemic syndrome)
(v) Renal tubule blockage
— e.g. 'myeloma kidney' (light chains)
3. Post-renal
— urinary tract obstruction

Acute-on-chronic renal failure

Acute deterioration of renal function in a patient with chronic renal failure may occur with dehydration or infection.

Features of acute renal failure
Oliguria or *anuria* are usually present in untreated ARF. The other features are as outlined already for renal failure in general.
Infection is a common cause of death in ARF.

Pre-renal ARF/acute tubular necrosis
— see Table 1 for the differentiation of these two
— in either case the patient should be rehydrated appropriately (if hypovolaemic) and given a diuretic (usually frusemide)
— in pre-renal ARF, if renal perfusion is restored, this will normally lead to immediate recovery of the renal failure
— in acute tubular necrosis, rehydration must be performed very carefully, because if it is overdone, the patient's kidneys will be unable to excrete the excess fluid/electrolytes and the circulation will be overloaded. Management should then be as described below for established ARP. Renal function will normally return within 3 weeks if pre-disposing causes have been removed and the patient can be kept alive.

Management of acute renal failure
Pre-renal failure should be treated as described above. Post-renal obstruction should be excluded and relieved if found. Otherwise, the management of established ARF is:
1. *Fluids*—maintain balance by giving a volume equivalent to the previous day's losses plus an allowance for insensible losses (insensible loss is normally about 500 ml/day). If the patient is fluid overloaded, the volume given should, of course, be reduced until the overload is corrected.*
2. *Sodium*—adjust sodium intake such that any overload or depletion is corrected,* then give an allowance equivalent to the previous day's losses. This will normally involve a low sodium diet.

Table 1 Differentiation of pre-renal failure from acute tubular necrosis

	Pre-renal ARF	*Acute tubular necrosis*
Urine osmolality	high	falls to a level similar to plasma
Urine: plasma osmolality ratio	> 1.5	< 1.1
Urine sodium	< 10 mmol/l	> 20 mmol/l
Response to diuretics	Yes	No

N.B. In pre-renal ARF the kidney is responding to poor renal perfusion by retaining sodium and water (→ low urine sodium and a concentrated urine with high osmolality)
As acute tubular necrosis develops, any glomerular filtrate produced passes out as urine with progressively less and less tubular modification, so the urine composition approaches that of plasma.

3. *Potassium*
   — low potassium diet
   — treat hyperkalaemia:*
   (i) acutely by giving glucose and insulin (which causes potassium to enter cells), or calcium gluconate (calcium antagonises the effect of hyperkalaemia on the heart)
   (ii) give calcium ion exchange resins orally or by enema (these remove potassium from the body and exchange it for calcium). Should be started at the same time as (i)
4. *Protein*—give low protein diet to reduce the production of nitrogenous waste products such as the 'middle molecules' which give rise to many of the symptoms of 'uraemia'.* Since the advent of renal dialysis, very low protein diets (20g/day) are no longer used because they increase the likelihood of infection.
5. *Acidosis*—a low protein diet will also reduce the production of acid and hence the severity of acidosis.* Do *not* treat with sodium bicarbonate as this would lead to sodium overload.
6. *Infection*—take all possible precautions to prevent infection and treat vigorously any which occurs.
7. *Drugs*—remember to adjust the dosage of drugs which are eliminated by the kidney.
8. *Diuretic phase of acute tubular necrosis*—at the time of recovery from acute tubular necrosis, the GFR usually starts to increase before tubular function has returned to normal. This leads to large volumes of dilute urine being passed which may contain considerable amounts of sodium and potassium. Careful monitoring of losses may be required at this stage.*

Chronic renal failure (CRF)

In chronic renal failure, the range of homeostatic adjustment of urine composition becomes progressively curtailed.

Initially, this has little effect except that the kidney is unable to concentrate urine as much as usual at night → nocturia.

Later, the kidney is unable to maintain homeostasis with the patient's existing dietary intake of fluid and electrolytes. In most cases, this is because the dietary load is too high for the failing kidney to be able to excrete it, the result being:

1. Sodium retention → hypertension and perhaps oedema
2. Water retention → hyponatraemia
3. Potassium retention → hyperkalaemia

(In *rare* cases, the ability of the kidney to retain sodium is more impaired than the ability to excrete, and then the normal dietary load of sodium may be insufficient to compensate for the urinary loss, resulting in a 'sodium-losing nephropathy'.)

The range over which the urine osmolality is adjustable becomes

* Renal dialysis may be required if these conservative measures fail.

smaller and smaller until eventually the kidney is only able to produce urine with a similar osmolality to plasma (i.e. *isosthenuria* is present).

These features of CRF can be explained by the *reduced nephron hypothesis*. As nephrons are progressively destroyed the plasma level of urea and other waste compounds rises. Thus the concentration of these compounds in the glomerular filtrate of the still-functioning nephrons also rises and this leads to an osmotic diuretic effect in the tubules.

Anaemia is a feature of chronic renal failure probably because of deficient erythropoietin production by the kidney.

The general biochemical features of chronic renal failure are as given on page 22.

Assessment of the rate of deterioration of renal function

— referring to Fig. 8 it will be seen that GFR is proportional to $\frac{1}{\text{plasma creatinine}}$

— GFR decreases linearly in chronic renal failure if there are no complicating factors

— a plot of $\frac{1}{\text{plasma creatinine}}$ will show a linear decline if there are no complicating factors.

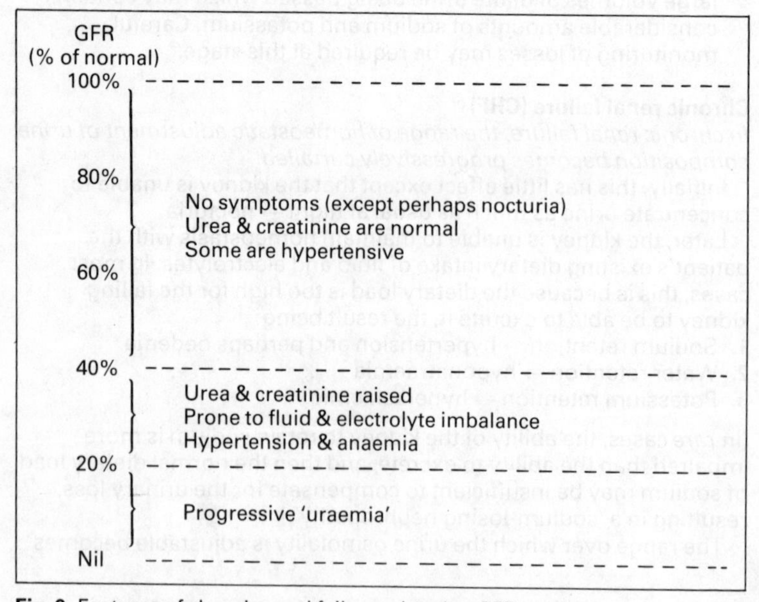

Fig. 9 Features of chronic renal failure related to GFR

Guidelines for management of chronic renal failure

1. Assess the rate of deterioration of renal function (see above)—if it accelerates, look for correctable factors
2. Control blood pressure carefully
3. Control water and salt balance—patients are at risk of fluid overload or dehydration, especially after intercurrent infections and surgery.
4. As necessary, modify doses of those drugs which are eliminated by the kidney
5. Control calcium and phosphate levels in an attempt to prevent renal osteodystrophy by giving:
   — phosphate binders
   — calcium supplements
   — vitamin D analogues (e.g. 1-alpha-OH cholecalciferol) which do not need to be hydroxlyated by the kidney
6. Reduce dietary protein intake—this appears to slow the progression of CRF as well as alleviating some complications
7. Renal transplantation or chronic dialysis will eventually be required

NEPHROTIC SYNDROME

Definition: Proteinuria sufficient to cause hypoalbuminaemia sufficient to cause oedema.

Normally proteinuria will need to exceed 0.05 g/kg body weight/day for the nephrotic syndrome to develop. Thus it will only occur with *glomerular disease* (see 'Urine Protein'—page 138).

Some causes of glomerular damage which may be sufficient to produce nephrotic syndrome:

1. Minimal change glomerulonephritis
   — this is a common cause of the nephrotic syndrome in children
2. Other idiopathic forms of glomerulonephritis
3. Connective tissue disorders
4. Infections (e.g. malaria, syphilis, hepatitis B)
5. Diabetes mellitus
6. Amyloidosis
7. Drugs (e.g. gold, penicillamine)

N.B. it should be stressed that nephrotic syndrome is, as its name implies, simply a *syndrome* associated with heavy proteinuria of any cause, and it should *not* be considered a final diagnosis in its own right

Investigation of nephrotic syndrome

Adults : a renal biopsy will normally be required

Children : if there is a highly selective proteinuria (see 'Selectivity of proteinuria', page 22) it is highly likely that minimal change glomerulonephritis is the cause and a renal biopsy can be avoided.

Associated biochemical findings in the nephrotic syndrome

1. Protein electrophoresis
   — albumin will, of course, be reduced
   — alpha-2 globulin increased (alpha-2 macroglobulin has a high molecular weight, so it is not readily leaked into the urine)
   — gamma globulin may be decreased if the proteinuria is non-selective
2. Secondary hyperlipidaemia (see page 153)

RENAL TUBULAR DISORDERS

N.B. Renal tubular disorders may occur as part of the general nephron damage occurring in renal failure. This section deals with those disorders in which there is only tubular dysfunction.

Classification

1. Multiple tubular disorders—'Fanconi syndrome' (see below)
2. Isolated tubular disorders
   (i) Renal glycosuria
   (ii) Renal tubular concentrating defects (see below)
   (iii) Renal tubular acidosis (see below)
   (iv) Aminoacidurias (see page 106)
   (v) Familial hypophosphataemic rickets (see page 95)
   (vi) Bartter's syndrome (see below)
   (vii) Pseudohypoparathyroidism (see page 93)

Fanconi syndrome

Characterised by:

1. glycosuria
2. phosphaturia
3. aminoaciduria
4. tubular proteinuria
5. hypouricaemia

Renal tubular acidosis and a defect in urinary concentrating ability usually occur as well. Sometimes urinary electrolyte loss may be a feature.

Causes:

1. Inborn errors of metabolism, e.g.
   (i) cystinosis
   (ii) glycogen storage disease } see chapter on Inborn Errors of Metabolism
   (iii) galactosaemia
   (iv) Wilson's disease (see page 126)
2. Acquired
   (i) exogenous toxic substance
      — especially heavy metals
      — some antibiotics
   (ii) multiple myeloma (*rare*)

Common presenting features:
1. Failure to thrive (in infancy)
2. Vitamin D resistant rickets
   — related to phosphate loss (and renal tubular acidosis if present)
3. Polyuria/polydipsia

Renal tubular concentrating defects
(see also 'Polyuria' page 15)

In these conditions there is renal resistance to the action of ADH
1. Inherited: nephrogenic diabetes insipidus (X-linked inheritance)
2. Acquired:
   (i) hypercalcaemia
   (ii) hypokalaemia (prolonged)
   (iii) drugs (e.g. lithium, demethylchlortetracycline)

Renal tubular acidosis (RTA)
Usually an inherited metabolic disorder but may be acquired (e.g. in chronic liver disease). Also occurs as part of the Fanconi syndrome.

Defect:
Related to failure of renal tubular hydrogen ion transport and hence failure of bicarbonate reabsorption. There are two types:
1. Distal RTA (or 'classical' RTA)
   — affecting the distal tubule
2. Proximal RTA
   — affecting the proximal tubule

Distal RTA
Distal tubular hydrogen ion transport has a *low* capacity but can normally work against a high pH gradient. In distal RTA, this transport mechanism is defective and hydrogen ion cannot be excreted against a significant pH gradient.

Features:
1. Urine pH always above 5.4
2. Chronic hyperchloraemic acidosis
3. Vitamin D resistant rickets caused by the chronic acidosis
4. Renal calculi and nephrocalcinosis
   — calcium salts precipitate out more readily in the constantly alkaline urine
5. Hypokalaemia
   — defective distal tubular H^+-Na^+ exchange leads to increased urinary sodium loss with a consequent secondary hyperaldosteronism → hypokalaemia

N.B. Hypokalaemia is usually associated with *alkalosis*. Hypokalaemic *acidosis* should arouse suspicion of a renal tubular disorder.

Tests:

1. early morning urine pH
   — normal people produce an acid urine at this time
   — a pH of < 5.4 excludes distal RTA
2. capillary (or arterial) blood pH and simultaneous urine pH
   — if the patient is acidotic and the urine pH is > 5.4 RTA is confirmed (though it could be either type of RTA)
3. urinary acidification test (see appendix)
   — should *not* be performed if the patient is acidotic
   — normal individuals should pass at least one urine specimen with a pH of < 5.4

Proximal RTA

Proximal tubular hydrogen ion transport has a *high* capacity but cannot function against a significant pH gradient. In proximal RTA this mechanism is defective, so proximal tubular bicarbonate reabsorption is impaired. Large amounts of bicarbonate pass on to the distal tubule where the *low* capacity distal transport mechanism is quickly saturated and bicarbonate is wasted in the urine.

An important difference from distal RTA is that an acid urine *can* be produced if the patient is sufficiently acidotic, because the filtered load of bicarbonate is then so low that the distal tubular transport mechanism is not saturated.

It is usually this form of RTA which occurs as part of the Fanconi syndrome.

The simplest way to differentiate proximal RTA from distal RTA is by the amount of alkali required to correct the acidosis. It is much greater in the proximal form.

Bartter's syndrome

— Recessively inherited
— Poorly understood but appears to be related to impaired proximal tubular sodium absorption. This leads to marked secondary hyperaldosteronism and hence the predominant feature is hypokalaemia.

RENAL CALCULI

Pathogenesis

1. Supersaturation of the urine with the crystalloid component of the stone caused by:
   (i) increased urinary excretion of the crystalloid
   (ii) variation of the pH of the urine resulting in diminished crystalloid solubility
      — calcium salts and magnesium ammonium phosphate are less soluble at alkaline pH
      — uric acid is less soluble at acid pH
   (iii) diminished urine volume
2. Deficiency of normal inhibitors of crystal formation

Types of renal stone disease

1. Calcium stone disease (approx. 80–85%)
   — dealt with in detail below
2. Infected stone disease (approx. 10%)
   — stones composed mainly of magnesium ammonium phosphate
   — main factors are high urine pH and ammonium concentration (urea-splitting organisms produce both of these)
3. Uric acid stone disease (approx. 5%)
   — main factors are low urine pH and increased urinary uric acid excretion
4. Cystine stone disease (approx. 1%)
   — see 'Cystinuria', page 105

Calcium stone disease

1. Majority of stones are calcium *oxalate* with or without calcium phosphate
2. Calcium phosphate content depends largely on the prevailing urinary pH
3. Causes:
   (i) Idiopathic calcium stone disease (see below)
   — approx. 85%
   (ii) Primary hyperparathyroidism (see page 91)
   — approx. 10%
   (iii) Renal tubular acidosis (see page 29)
   (iv) Gross hyperoxaluria
   — Hereditary hyperoxaluria
   — Enteric hyperoxaluria: intestinal oxalate absorption is increased—usually associated with Crohn's disease or small bowel resection.
   (v) Miscellaneous
   — vitamin D intoxication
   — sarcoidosis
   — milk-alkali syndrome
   — immobilisation
   — Cushing's syndrome and steroid treatment

Idiopathic calcium stone disease

1. Main metabolic abnormality is probably increased absorption of calcium from the gut—probably related to increased plasma 1,25-dihydroxycholecalciferol possibly stimulated by:
2. Hypophosphataemia (30% of cases)
3. Normal serum calcium and parathyroid hormone levels
4. Mildly *increased urinary oxalate excretion* is probably important in the stone formation. In the gut, dietary oxalate combines with calcium to form insoluble and poorly absorbed calcium oxalate. If the intestinal content of calcium is reduced (by hyperabsorption of calcium or reduced dietary intake) more dietary oxalate is absorbed and then excreted in the urine.

5. Hypercalciuria is present in 15–25% of patients—the term 'idiopathic hypercalciuria' has been used to describe this condition.
6. Management:
   (i) give high fluid intake
   (ii) correct any dietary excesses of oxalate or calcium
   If necessary proceed to:
   (iii) thiazide diuretics
   — hypocalciuric action
   — effective even in the absence of hypercalciuria
   or:
   (iv) oral phosphate supplements
   — reduce calcium absorption

DIURETICS

Osmotic diuretics

Osmotic diuretics are low molecular weight substances that are freely filtered by the glomerulus and remain in the tubular lumen in high concentration because of a limitation on their reabsorption. Thus they contribute notably to the osmolality of the filtrate and so inhibit the reabsorption of water, sodium and other electrolytes.

Examples of osmotic diuresis:
1. Hyperglycaemia
   — when the tubular maximum reabsorptive capacity for glucose is exceeded—usually at plasma glucose levels above about 10 mmol/l
2. Chronic renal failure
   — see 'reduced nephron hypothesis', page 26
3. Osmotic diuretic drugs
   — e.g. mannitol

Loop diuretics

(frusemide, ethacrynic acid and bumetanide)
— these are potent diuretics
— they act mainly by inhibiting sodium chloride reabsorption in the medullary part of the ascending limb of the loop of Henle.

Thiazide diuretics

— these are moderately potent diuretics
— they act by inhibiting sodium chloride reabsorption in the cortical part of the ascending limb of the loop of Henle (sometimes called the 'distal diluting segment' as it is this part of the loop which produces a hypotonic tubular fluid)

Potassium-sparing diuretics
(spironolactone, amiloride, triamterene)

— these act on the distal tubule and inhibit the reabsorption of sodium in exchange for potassium/hydrogen ion—hence the potassium-sparing effect

— spironolactone has this action because it inhibits aldosterone, the others act independently of aldosterone

N.B. the loop and thiazide diuretics all tend to produce *hypokalaemic alkalosis*. This is because they increase sodium delivery to the distal tubule so that more is available here for exchange with potassium/hydrogen ion.

Acid-base balance/oxygen carriage

PHYSIOLOGY

Control of pH is by means of:

1. Buffers
   - (i) Bicarbonate
   - (ii) Protein
   - (iii) Phosphate
2. Respiratory compensation
3. Renal compensation

Bicarbonate buffer system

This is particularly important because two components can be regulated:

$$CO_2 + H_2O \rightleftharpoons H_2CO_3 \rightleftharpoons H^+ + HCO_3^-$$

$CO_2 + H_2O$ ↑ regulated by the lungs; $H^+ + HCO_3^-$ ↑ regulated by the kidneys

The Henderson-Hasselbalch equation will be mentioned here for completeness though it will not be used in this chapter for any of the explanations of acid-base disturbances. It is:

$$pH = pK' + \log \frac{[HCO_3^-]}{[CO_2 \text{ dissolved}]} \qquad \text{where } pK' = 6.1$$

If S is the solubility coefficient of CO_2:

$$[CO_2 \text{ dissolved}] = S \times P\text{co}_2$$

$$pH = pK' + \log \frac{[HCO_3^-]}{S \times P\text{co}_2}$$

Thus pH is dependent on the ratio $\frac{[HCO_3^-]}{S \times P\text{co}_2}$

Both components of this ratio can be regulated.

DISTURBANCES OF ACID-BASE BALANCE

Acidaemia—decrease in pH (increase in $[H^+]$)
Alkalaemia—increase in pH (decrease in $[H^+]$)

Acidosis / Alkalosis } Strictly these describe abnormal processes or conditions which would cause a deviation of pH if there were no secondary compensatory changes (see below) but in practice, they are commonly used as if they were synonymous with acidaemia and alkalaemia respectively.

1. Respiratory disturbances
   (i) Acidosis
   impaired respiratory elimination of CO_2
   → ↑ $P\text{CO}_2$ → acidaemia
   (ii) Alkalosis
   increased respiratory elimination of CO_2
   → ↓ $P\text{CO}_2$ → alkalaemia
2. Metabolic disturbances
   (i) Acidosis
   accumulation of *fixed* acid (i.e. any acid except carbonic acid) and/or loss of base
   (ii) Alkalosis
   accumulation of base or loss of fixed acid
3. Mixed disturbances
   Occasionally primary metabolic and respiratory disturbances may occur simultaneously.

Compensation

In *chronic* respiratory disturbances, renal adjustment of the plasma $[HCO_3^-]$ will usually occur to bring the pH back towards normal.

In metabolic disturbances (unless of very short duration) respiratory adjustment of the $P\text{CO}_2$ will usually occur to bring the pH back towards normal. (If the kidneys are functioning normally, they will then excrete acid or base in an attempt to correct the metabolic disturbance itself).

Causes of acid-base disturbances

1. Respiratory acidosis
   (i) Lung disease e.g. chronic bronchitis and emphysema, pulmonary oedema, advanced T. B., pneumonia, pulmonary fibrosis, respiratory distress syndrome
   (ii) Airway obstruction and suffocation
   (iii) Disorders of the thoracic cage e.g. trauma, kyphoscoliosis
   (iv) Muscle and nerve disorders affecting respiratory movements e.g. polio, high spinal injuries
   (v) Respiratory centre disturbances e.g. drugs (especially in overdose), cerebro-vascular accidents

2. Respiratory alkalosis
   (i) Stimulation of the respiratory centre e.g. anxiety, hypoxia, salicylates
   (ii) Mechanical ventilation (not properly controlled)
3. Metabolic acidosis
   (i) With increased anion-gap (see below)
      a. Diabetic ketoacidosis
      b. Renal failure
      c. Lactic acidosis (see below)
      d. Some overdoses/poisonings e.g. salicylates, methanol, ethylene glycol (anti-freeze)
   (ii) Hyperchloraemic acidosis (normal anion-gap)
      a. Renal tubular acidosis (see page 29)
      b. Others e.g. high intestinal fistulae, very severe diarrhoea, therapy with carbonic anhydrase inhibitors (acetazolamide), following implantation of the ureter into the intestine.
4. Metabolic alkalosis
   (i) Increased intake of base e.g. sodium bicarbonate for indigestion
   (ii) Increased loss of acid
      a. from the stomach e.g. pyloric stenosis, prolonged aspiration
      b. renal e.g. prolonged diuretic therapy, Cushing's syndrome

Anion-gap

This is calculated from the concentration of electrolytes in plasma thus:

Anion gap = $([Na^+] + [K^+]) - ([Cl^-] + [HCO_3^-])$

An anion gap of greater than 20 mmol/l is almost certainly pathological and demonstrates the build up in plasma of an abnormal concentration of an anion other than chloride or bicarbonate:

Diabetic ketoacidosis	: ketoacid anions	
Renal failure	: a range of acid anions	
Lactic acidosis	: lactate	
Methanol poisoning	: formate	produced by metabolism of the poison
Ethylene glycol poisoning	: oxalate	

Lactic acidosis

Causes:

1. Tissue hypoxia—by far the most common cause
   (i) Hypoxaemia
   (ii) Hypovolaemia

2. Diabetes (see page 65)—usually associated with phenformin therapy
3. Idiopathic
4. Some inherited metabolic disorders e.g. glycogen storage disease type I

LABORATORY ASSESSMENT

Sample requirements

For complete elucidation of acid-base status, *anaerobic* samples are essential. Ideally, *arterial* blood is taken into a heparinised syringe (ensuring that no air bubbles are included) and the syringe is capped immediately. The results on carefully collected capillary blood agree well with arterial blood (except perhaps for Po_2) unless there is impaired peripheral blood flow.

Sample stability

— samples must be analysed within 20 minutes unless cooled to 4°C when they can generally be stored for an hour or two.

N.B. Samples with a high leukocyte count will deteriorate more rapidly and the only safe solution is to analyse all samples immediately.

Parameters measured or calculated

1. pH
   Measured directly using a specially designed pH electrode
   N.B. some labs. report hydrogen ion concentration instead
2. Pco_2
   This is a *good* measure of respiratory disturbance and is *not* affected by metabolic disturbances
   N.B. Metabolic disturbances might have been expected to affect the Pco_2 e.g. in a metabolic acidosis, the addition of fixed acid will shift the following equilibrium to the left by increasing the hydrogen ion concentration:
   $$H_2O + CO_2 \rightleftharpoons H_2CO_3 \rightleftharpoons H^+ + HCO_3^-$$
   Thus the Pco_2 will tend to rise, but on passage through the lungs, the extra CO_2 will be lost and so in arterial blood, the Pco_2 will be unchanged.
3. Metabolic component
   (i) Bicarbonate (i.e. *actual bicarbonate*)
   Not ideal for assessment of metabolic disturbances because it is affected by respiratory disturbances e.g. in a respiratory acidosis, the bicarbonate concentration will rise because the following reaction will be shifted to the right by the increased Pco_2:
   $$H_2O + CO_2 \rightleftharpoons H_2CO_3 \rightleftharpoons H^+ + HCO_3^-$$

(ii) Standard bicarbonate
This is the bicarbonate concentration of fully oxygenated blood at 37°C after equilibration of the whole blood at a Pco_2 of 40 mmHg (5.3 kPa). Thus any change in the bicarbonate concentration caused by a change in Pco_2 (i.e. caused by a respiratory disturbance) will very largely be removed, so standard bicarbonate is a *good* measure of metabolic disturbance.

(iii) Base excess
This is the concentration of base in whole blood as measured by titration with strong acid to pH 7.4 at a Pco_2 of 40 mmHg (5.3 kPa) at 37°C. For negative values (i.e. base deficit) the titration is carried out with strong base.
As for standard bicarbonate, the equilibration to normal Pco_2 makes this a *good* measure of metabolic disturbance (though expressed in a different way).

(iv) E. C. F. base excess (or standard base excess)
(ii) and (iii) are based on the assumption that the in vitro CO_2 titration line for whole blood is identical to the in vivo CO_2 titration line. However, in vivo, the whole extracellular fluid is involved in buffering, and its buffering capacity approximates that of blood in vitro with a haemoglobin of 6 g/dl. This discrepancy may lead to misinterpretation of the acid-base status e.g. in an acute pure respiratory acidosis there may appear to be a slight metabolic acidosis in addition.
The E. C. F. base excess is based on in vivo CO_2 titration findings and is thus the *best* measurement of metabolic component. It is a derived value and cannot be measured directly.

N.B. all these estimates of metabolic component are normally derived from the pH and Pco_2 by means of a nomogram.

Nomograms

Several are available, but the most useful are:

1. Siggaard Andersen alignment nomogram (see Fig. 10)
2. New Siggaard Andersen nomogram (see Fig. 11)

The data plotted in Figure 10 correspond to a pH of 7.12, Pco_2 of 90 mmHg and haemoglobin of 15 g/dl. A line is drawn through the corresponding points on the pH and Pco_2 scales. The actual bicarbonate can then be read off from where this line crosses the HCO_3^- scale. The base excess can be read off from the base excess scale corresponding to the patient's haemoglobin, while the E. C. F. base excess is obtained from the scale corresponding to a haemoglobin of 6 g/dl. To obtain the standard bicarbonate, a line is drawn through the point corresponding to the base excess and the

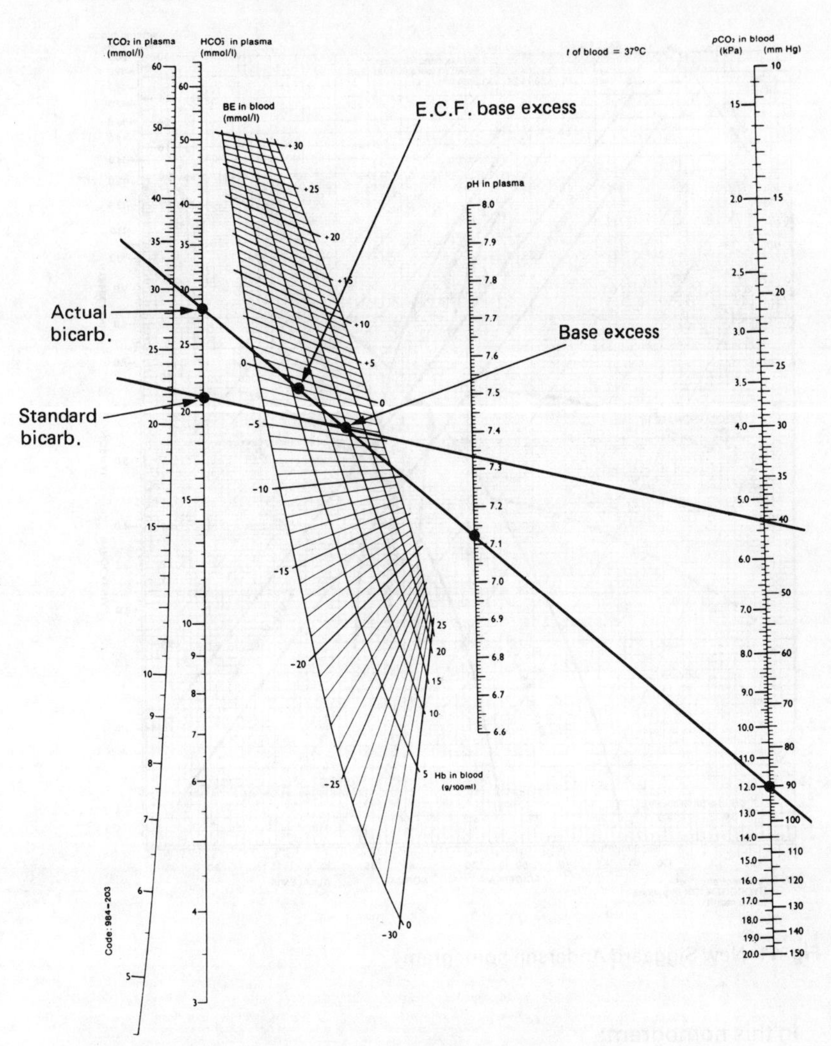

Fig. 10 Siggaard Andersen alignment nomogram

40 mmHg mark on the Pco_2 scale. The intersection of this line with the HCO_3^- scale gives the standard bicarbonate.

In Figure 11, the point corresponding to the patient's Pco_2 and pH (or hydrogen ion concentration) will generally fall within one of the areas marked as corresponding to one of the various acid-base states. The ECF base excess can be read from the diagonal scale.

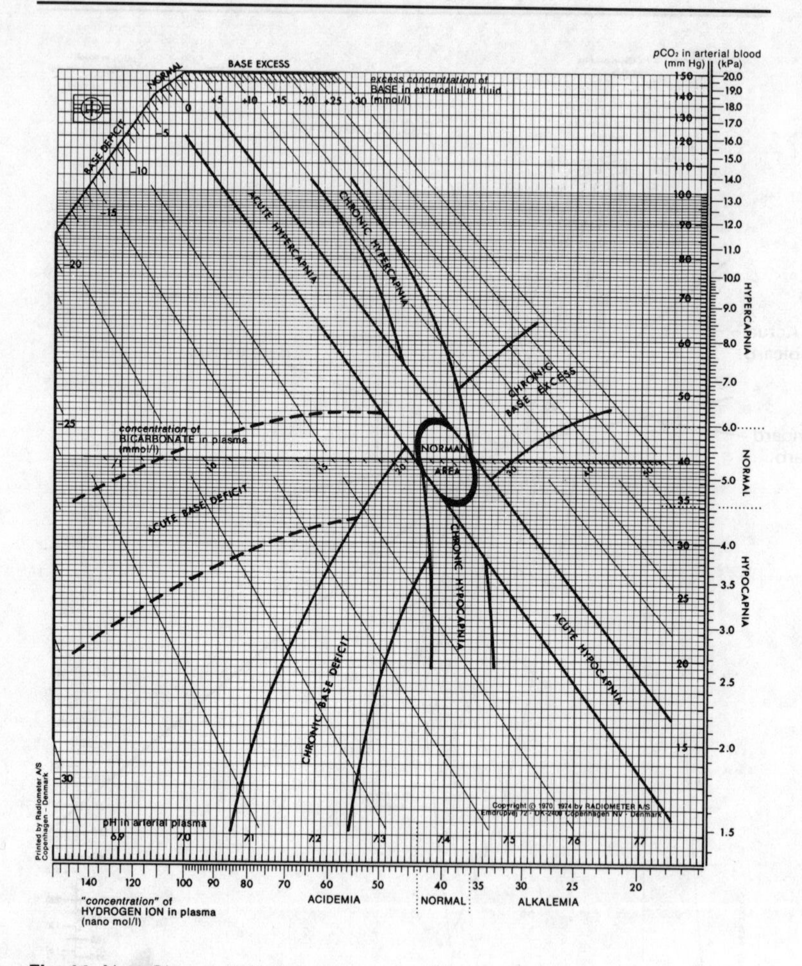

Fig. 11 New Siggaard Andersen nomogram

In this nomogram:

acute	≡	uncompensated
chronic	≡	compensated
hypercapnia	≡	respiratory acidosis
hypocapnia	≡	respiratory alkalosis
base excess	≡	metabolic alkalosis
base deficit	≡	metabolic acidosis

Temperature corrections
pH and $P\text{co}_2$ are usually determined at 37°C. If the patient's temperature differs significantly from this, corrections must be applied.

Interpretation of acid-base results

		pH	*PCO_2*	*Std. bicarb.*	*Base excess*
Respiratory acidosis	(uncompensated)	↓	↑	N	N
" "	(compensated)	↓	↑	↑	+ve
Respiratory alkalosis	(uncompensated)	↑	↓	N	N
" "	(compensated)	↑	↓	↓	−ve
Metabolic acidosis	(uncompensated)*	↓	N	↓	−ve
" "	(compensated)	↓	↓	↓	−ve
Metabolic alkalosis	(uncompensated)*	↑	N	↑	+ve
" "	(compensated)	↑	↑	↑	+ve
Mixed metabolic & respiratory acidosis		↓	↑	↓	−ve

* Respiratory compensation usually occurs quickly so these are rarely seen

OXYGEN CARRIAGE

Most oxygen is carried in the blood bound to haemoglobin. The percentage saturation of haemoglobin with oxygen is a function of $P\text{o}_2$ and follows the familiar sigmoidal curve (Fig. 12).

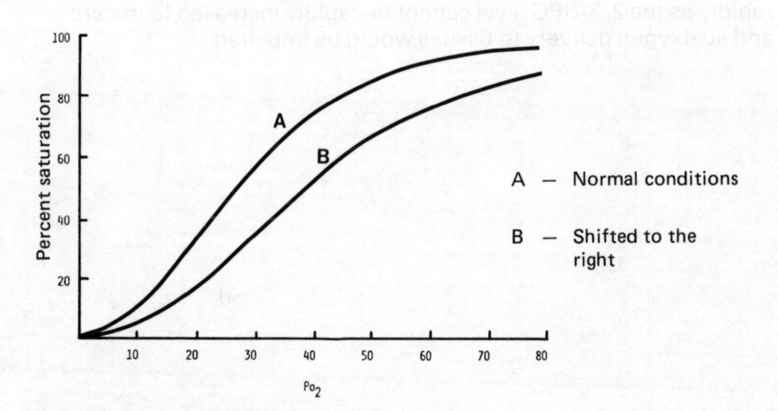

Fig. 12 Oxyhaemoglobin dissociation curve

Factors shifting the oxyhaemoglobin dissociation curve

To the right:

1. increase in hydrogen ion concentration (decrease in pH) i.e. the Bohr–Haldane effect.
2. increase in temperature
3. increase in Pco_2
4. increase in erythrocyte 2, 3-diphosphoglycerate (2, 3-DPG) concentration

To the left:

1. increase in carbon monoxide concentration
2. the presence of haemoglobin F

N.B. A shift to the *right* represents a decreased affinity of haemoglobin for oxygen—at normal physiological oxygen tensions, more oxygen will be delivered to the tissues.

Oxygen saturation

This is often calculated from measured Po_2 and pH. In this case the other factors affecting the true oxygen saturation are ignored and misleading results can be obtained.

2, 3-diphosphoglycerate (2, 3-DPG)

Control of the erythrocyte 2, 3-DPG concentration appears to be a mechanism whereby oxygen delivery to tissues can be adjusted to compensate for a shift in the oxyhaemoglobin dissociation curve—e.g. in chronic acidosis, the erythrocyte 2, 3-DPG level is reduced. Thus, in such situations, it is not advisable to correct the acidosis rapidly as the 2, 3-DPG level cannot be rapidly increased to match, and so oxygen delivery to tissues would be impaired.

Liver

PHYSIOLOGY

Functions of the liver

1. Metabolic functions
   (i) Carbohydrate metabolism
      a. glycogen synthesis, glycogen storage and glycogenolysis
      b. gluconeogenesis
   (ii) Protein metabolism
      a. deamination of amino acids and synthesis of urea from the amino groups
      b. synthesis of most plasma proteins except the immunoglobulins
   (iii) Lipid metabolism
      a. oxidation of fatty acids to ketone bodies
      b. synthesis of triglyceride, cholesterol and phospholipids
      c. role in the metabolism of lipoproteins (see Chapter XVI)
   (iv) Bilirubin metabolism (see below)
   (v) Vitamin metabolism
   (vi) Inactivation of many hormones
2. Excretory/digestive functions
   (i) bile acid production
      — bile acids are formed from cholesterol solely in the liver
      — bile acids facilitate the emulsification of lipids in the intestine prior to digestion
   (ii) bilirubin excretion (see below)
   (iii) excretion of relatively non-polar compounds e.g. cholesterol
3. Detoxication of many drugs, poisons etc.
4. Haematological/immunological/reticulo-endothelial/circulatory functions

Bilirubin metabolism/excretion

1. Haemoglobin is broken down in reticulo-endothelial cells, bilirubin being derived from the haem moiety after removal of iron.
2. This *unconjugated* bilirubin is transported to the liver bound to plasma albumin—as a consequence of protein-binding, it is not

filtered by the renal glomerulus and so does not pass into the urine.

3. Bilirubin is excreted in the bile by a 3 stage process in the liver:
   (i) uptake into the liver cell by an active process
   (ii) conjugation with glucuronic acid by the activity of the enzyme glucuronyl transferase
   (iii) transfer out of the liver cell into the biliary canaliculus
4. In the intestine, conjugated bilirubin is metabolised by bacteria to stercobilinogen. The bulk of this is excreted in the stool, but some is reabsorbed. Most of the latter is re-excreted by the liver into the bile, but a small amount passes into the systemic circulation and is then excreted in the urine (where it is called *urobilinogen*).

LIVER FUNCTION TESTS (LFT'S)

Routine biochemical LFT's

Serum

1. Bilirubin (see 'Jaundice' below)
   — conjugated bilirubin (also called 'direct bilirubin') and unconjugated ('indirect') bilirubin can be measured separately, though this is not usually done routinely.
2. Alkaline phosphatase (see page 161)
   — produced by the cells of bile canaliculi.
   — elevated levels occur with *biliary obstruction* as a result of the associated proliferation of bile canaliculi.
   — in the presence of jaundice, levels over two and a half times the upper limit of normal indicate undoubted obstructive (cholestatic) jaundice.
   — smaller elevations occur with mild obstruction and hepato-cellular disease (in the latter this may be related to mild obstruction caused by swelling of the liver cells).
   — remember that the reference interval is higher in children than in adults because of bone growth.
3. Aminotransferases i.e. 'transaminases' (see page 160)
   — Aspartate aminotransferase (AST) and alanine aminotransferase (ALT)
   — released from liver cells with *hepato-cellular damage*; in liver disease very high levels are diagnostic of this.
   — smaller elevations occur with mild hepato-cellular damage or with cholestasis.
   — elevations of ALT are fairly specific for liver disease but remember that AST also rises with myocardial damage and skeletal muscle damage (though the origin of an elevated AST is usually obvious clinically).
   — generally little extra information is gained from estimating both AST and ALT, so most laboratories only measure one

routinely (this is usually AST as it is useful in conditions other than liver disease and it also tends to be more sensitive in detecting chronic liver disease).

4. Albumin and total protein
   — reduced synthesis of plasma proteins occurs with many liver diseases but is not specific
   — increased synthesis of some immunoglobulins occurs with many chronic liver diseases and the pattern of increase may provide supportive evidence towards a diagnosis (see 'Immunoglobulins'—page (134).
5. γ –glutamyl transferase (see below)
   — some laboratories now include this as a routine LFT but for most it is a second line test.

Urine
Screening tests for:

1. Bilirubin
   — normally absent
   — absent in unconjugated hyperbilirubinaemia (see 'Jaundice' below)

   N.B. dip-stick tests are not sensitive enough to exclude the presence of small amounts of bilirubin in the urine.
   — present with other causes of jaundice.
2. Urobilinogen
   — not a very useful test
   — normally present in fresh urine.
   — absent in the presence of complete biliary obstruction.

Other LFT's

Biochemical

1. Serum γ –glutamyl transferase (γ GT) (see page 161)
   — probably best considered to be a sensitive test for biliary obstruction, but it is also elevated with hepato-cellular damage.
   — usually the first serum enzyme to rise with hepatic secondaries, with other space occupying lesions in the liver and with chronic liver disease (e.g. cirrhosis)
   — elevated serum levels also occur with chronic exposure to hepatic microsomal enzyme inducing agents (especially *alcohol* and drugs like phenobarbitone, phenytoin and rifampicin)—this may cause confusion if not remembered.
2. Serum 5′–nucleotidase (5NT) (see page 161)
   — difficult to measure reliably so not performed by many laboratories.
   — elevated specifically with biliary obstruction.
   — unlike γ GT it is not affected by enzyme inducing agents.

3. Alkaline phosphatase isoenzymes (see page ***)
4. Serum protein electrophoresis and immunoglobulins
   — (see 'albumin and total protein' above)
5. Bromsulphthalein (BSP) excretion (see Appendix for details)
   — a very sensitive test of liver disease
   — reflux of BSP from the liver cell into the circulation 90 mins after the injection is diagnostic of Dubin-Johnson syndrome (see 'Jaundice' below).
6. Serum bile acids
   — very sensitive test of hepatobiliary disease (especially when measured 2 hours after a meal)
   — technically difficult to measure
   — not generally available
7. Tests for specific disorders affecting the liver
   (i) Wilson's disease (see page 126)
      — serum copper and caeruloplasmin
      — urinary copper
   (ii) Haemochromatosis (see page 120)
      — serum iron and total iron binding capacity
      — serum ferritin
   (iii) alpha-1-antitrypsin deficiency (see page 137)
      — serum alpha-1-antitrypsin
   (iv) primary liver cell cancer (see page 141)
      — alpha-fetoprotein

Other pathology departments

1. Full blood count and blood film report
   — may suggest:
      (i) Liver disease (target cells, spur cells)
      (ii) Alcohol abuse (high mean corpuscular volume)
      (iii) Haemolysis
2. Prothrombin time (PT) and partial thromboplastin time (PTT)
   — most clotting factors (except VIII) are synthesised by the liver
   — Synthesis of II, VII and IX are vitamin K dependent and absorption of the latter is impaired in the presence of biliary obstruction (it being a fat-soluble vitamin).
   — both tests depend on the presence of I, II and X. PT also tests the extrinsic pathway (involving VIII); PTT the intrinsic pathway (VIII, IX, XI and XII).
3. Antibody titres
   (i) Mitochondrial antibody (primary biliary cirrhosis)
   (ii) Antinuclear factor } chronic active hepatitis
   (iii) Smooth muscle antibody } chronic active hepatitis
4. Hepatitis B antigens
   — for the diagnosis of hepatitis B infections

JAUNDICE

1. Pre-hepatic
   - (i) Increased production of unconjugated bilirubin
     - a. haemolysis
     - b. ineffective erythropoiesis
2. Hepatic
   - (i) Physiological jaundice in neonates
   - (ii) Acute hepato-cellular damage
     - a. viral hepatitis
     - b. toxic liver damage
   - (iii) Chronic hepatocellular damage
     - a. cirrhosis
     - b. chronic active hepatitis
   - (iv) Intra-hepatic cholestasis
     - a. space occupying lesions (especially secondary deposits)
     - b. drugs (e.g. chlorpromazine)
     - c. primary biliary cirrhosis
   - (v) Inherited disorders of bilirubin metabolism
     - a. Gilbert's syndrome
     - b. glucuronyl transferase deficiency
     - c. Rotor/Dubin-Johnson syndromes

     } see below
3. Post-hepatic (obstruction/compression of the bile ducts)
   - (i) Gall stones
   - (ii) Carcinoma of the head of the pancreas
   - (iii) Enlarged lymph nodes at the porta hepatis
   - (iv) Biliary tract neoplasms
   - (v) Bile duct stenosis (e.g. post-surgery) or atresia

Inherited disorders of bilirubin metabolism

1. Gilbert's syndrome
   - — common
   - — impaired uptake of bilirubin into the liver cell
   - — elevated serum level of *unconjugated* bilirubin but usually less than 50 micromol/l
   - — otherwise normal liver function tests including γ GT and usually normal BSP excretion
   - — no bilirubin in the urine
   - — needs to be excluded as a cause of hyperbilirubinaemia but otherwise of no great significance—indeed some believe that it is simply a reflection of the fact that serum bilirubin has a skewed distribution in the normal population with a tail of fairly high levels and that Gilbert's simply represents the upper end of that tail.
2. Glucuronyl transferase deficiency
   - — rare
   - — elevated serum level of *unconjugated* bilirubin
   - — no bilirubin in the urine

Table 2 Typical biochemical findings in some liver disorders

DISORDER	UNCONJ. BILI.	CONJ. BILI.	URINE BILI.	ALK. PHOS.	AST	ALBUMIN	γGT	BSP
Physiological neonatal jaundice	↑	N	Neg	N	N	N	N	*
Haemolysis/ ineffective erythropoiesis	↑	N	Neg	N	N	N	N	N
Gilbert's syndrome	↑	N	Neg	N	N	N	N	N
Hepatitis (early)	↑↑	↑↑	Pos	N or ↑	↑↑↑	N	↑↑	*
(cholestatic phase)	↑	↑↑↑	Pos	↑↑	↑↑	N	↑↑	*
Cirrhosis (early)	N	N	Neg	N	N	N	↑	Abn.
(late)	↑↑	↑↑	Pos	↑↑	↑↑	Low	↑↑	*
Obstructive jaundice	↑	↑↑↑	Pos	↑↑↑	N or ↑	N	↑↑↑	*

* the BSP test is not indicated in these conditions—it does not aid the diagnosis and may be associated with a high incidence of adverse reactions.

— mild and severe forms
— both usually present in neonates
— in the severe form (known as Crigler-Najjar syndrome) kernictus is almost inevitable

3. Dubin-Johnson and Rotor syndromes
   — both rare
   — failure of excretion of conjugated bilirubin from the liver cell into the bile canaliculus → *conjugated* hyperbilirubinaemia and bilirubin in the urine.
   — in Dubin-Johnson syndrome, a black pigment collects in the liver cell
   — in Dubin-Johnson syndrome, though BSP excretion is normal at 45 mins., BSP cannot be excreted from the liver cell into the bile canaliculus and so it refluxes into the circulation giving rise to a higher level at 90 mins. than at 45 mins.

Elucidation of the cause of jaundice

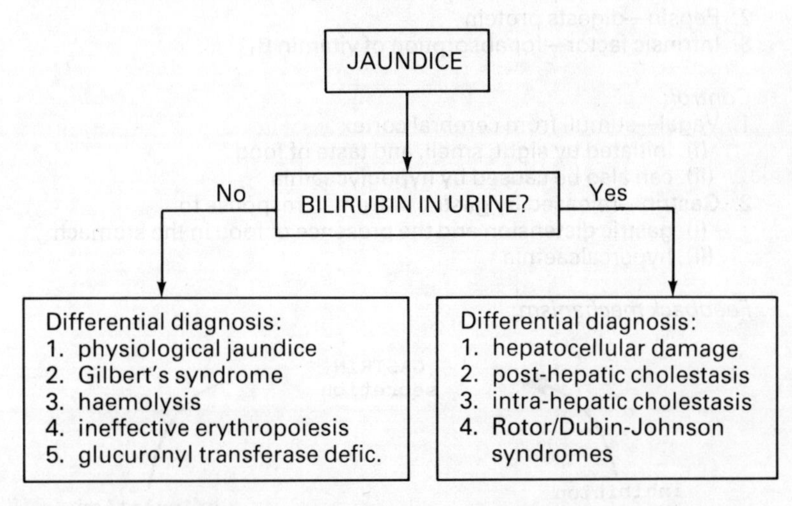

Other liver function tests may be of help in the further elucidation of the cause of the jaundice (see "LFT's" above and Table 2), but in many cases biochemical tests are not diagnostic.

Gastrointestinal tract

PHYSIOLOGY

Gastric secretion

Function:

1. Acid—initiates protein digestion
2. Pepsin—digests protein
3. Intrinsic factor—for absorption of vitamin B_{12}

Control:

1. Vagal—stimuli from cerebral cortex
   (i) initiated by sight, smell, and taste of food
   (ii) can also be caused by hypoglycaemia
2. Gastrin—released by gastric antrum in response to
   (i) gastric distension and the presence of food in the stomach
   (ii) hypercalcaemia

Feedback mechanism:

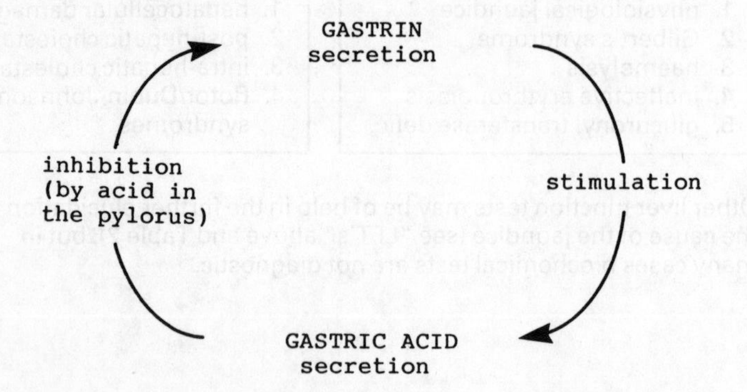

Pancreatic secretion

The pancreas is responsible for the secretion of most of the important digestive enzymes but 90% of the pancreas may be removed before protein digestion is impaired.

1. Bicarbonate
   Function: neutralisation of gastric acid to achieve optimum pH for pancreatic enzyme activity.
   Control: secretin—released in response to acid and food entering the duodenum.
2. Proteolytic enzymes

(i) trypsinogen	→ trypsin	activated in the duodenum
(ii) chymotrypsinogen	→ chymotrypsin	
(iii) proelastase	→ elastase	
(iv) procarboxypeptidase	→ carboxypeptidase	

   Function: protein breakdown to oligopeptides and aminoacids
   Control: pancreozymin (also called cholecystokinin)—released in response to acid and food entering the duodenum

N.B. Vagal stimulation and cholinergic agents potentiate the effects of secretin and pancreozymin BUT vagotomy does not result in clinically recognisable impairment of digestion.

3. Amylase
   Function: hydrolyses polysaccharides (starch and glycogen) to disaccharides (maltose, sucrose, lactose)
   Control: pancreozymin (see 2. above)
4. Enzymes for digestion of fats
   Function:
   (i) lipase
   — hydrolyses triglyceride to fatty acids and monoglyceride
   (ii) colipase
   — counteracts the inhibition of pancreatic lipase by bile salts
   Control: pancreozymin (see 2. above)

Biliary secretion

Bile salts are the major digestive components of bile
Function: to emulsify fat entering the duodenum (and so aid digestion by lipase etc.)
Control: gall bladder contraction stimulated by cholecystokinin (also called pancreozymin—see pancreatic secretion above)

N.B. enterohepatic circulation of bile salts (bile salts are reabsorbed in the small intestine and reused)

Normal intestinal absorption

1. Protein absorption depends on
   (i) normal pancreatic amylase (polysaccharides)
   (ii) normal intestinal mucosa—active transport mechanisms
2. Carbohydrate absorption depends on
   (i) normal pancreatic amylase (polysaccharides)
   (ii) intestinal disaccharidases (disaccharides)
   (iii) normal intestinal mucosa—active transport mechanisms (monosaccharides)

3. Fat absorption depends on
   (i) bile salts—to emulsify fats
   (ii) normal pancreatic lipase—to digest triglycerides
   (iii) normal intestinal mucosa—for chylomicron formation
4. Calcium absorption depends on
   (i) low level of intestinal phosphate and fatty acids (these form insoluble salts with calcium)
   (ii) presence of vitamin D (which in turn depends on normal fat absorption since it is a fat soluble vitamin)
   (iii) normal intestinal mucosa
5. Iron absorption depends on
   (i) iron being in the ferrous form
   (ii) active transport mechanism
6. Vitamin B_{12} absorption depends on
   (i) normal intrinsic factor
   (ii) intact mucosa in distal ileum
   (iii) normal intestinal flora (some bacteria compete for B_{12})

Major sites of absorption in the small intestine

Proximal	*Mid*	*Distal*
iron	aminoacids	bile salts
calcium	(monosaccharides)	vitamin B_{12}
water-soluble vitamins		(aminoacids)
monoglycerides		
fatty acids		
monosaccharides		
(aminoacids)		

G. I. hormones

A diffuse endocrine system exists throughout the gastrointestinal tract producing a variety of hormones which play an important part in the control of digestion. Some of these have already been mentioned but in man the role of the others is, in many instances, not very clear.

Location and proposed action of the best known G. I. hormones

Gastrin	antrum of the stomach and proximal small intestine	stimulates gastric acid secretion
Secretin	duodenum and jejunum	stimulates secretion of bicarbonate-rich pancreatic juice
Glucagon	pancreas	insulin antagonist
Pancreatic polypeptide	pancreas	inhibits pancreatic enzyme secretion

Pancreozymin (cholecystokinin)	small intestine	stimulates pancreatic enzyme secretion and gall bladder contraction
Gastric inhibitory polypeptide (GIP)*	small intestine	stimulates insulin release in response to oral glucose
Vasoactive intestinal polypeptide (VIP)	small intestine	increases intestinal motility

* In view of what is now believed to be its main role, it has been suggested that GIP should be renamed '*G*lucose-dependent *I*nsulin-releasing *P*olypeptide'

THE PANCREAS

Assessment of pancreatic function

1. Serum enzyme estimations
   (i) amylase (urine amylase may also be measured)
   (ii) trypsin
   (iii) lipase
2. Function tests (see below)
   (i) exocrine
      a. secretin test (measure of bicarbonate output)
      b. Lundh test (pancreatic digestive capacity)
         — dependent on an intact small gut mucosa
      c. PABA test (of chymotrypsin hydrolysis)
         — screening test only
      d. faecal fat excretion
   (ii) endocrine
      — glucose tolerance test

In general, enzyme measurements are of most value in acute pancreatitis and function tests in chronic pancreatic disease. Anatomical visualisation may help with either.

Acute pancreatitis

Aetiology:
Not fully understood but commonly associated with:
1. alcoholism
2. biliary tract disease

May be precipitated by:
1. hypercalcaemia (N.B. hyperparathyroidism)
2. hyperchylomicronaemia

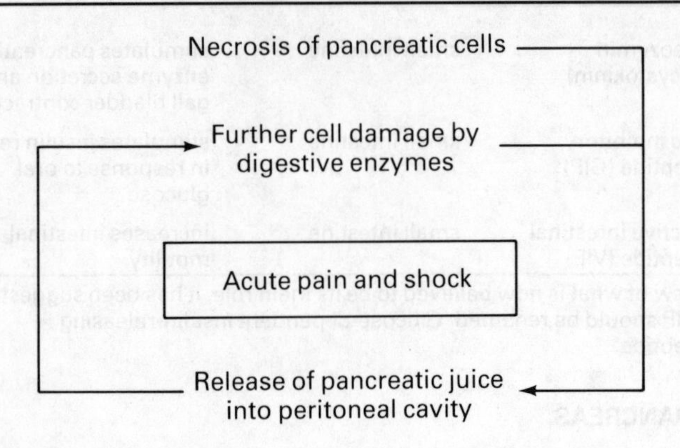

Laboratory investigation:
There is no diagnostic test for acute pancreatitis
1. Serum amylase
   — classically elevated for about 12–100 hours from the onset of an attack *but* may also be increased in:
     (i) mumps (salivary amylase)
     (ii) almost any acute inflammatory condition of the upper abdomen e.g. cholecystitis, perforated peptic ulcer.
     (iii) ruptured ectopic pregnancy
     (iv) renal failure (failure of amylase excretion)
   — persistent elevation suggests
     (i) development of pancreatic pseudocyst
     (ii) macroamylasaemia (uncommon but confusing)

N.B. Serum amylase *may* be normal or only slightly elevated in acute pancreatitis

2. Urine amylase
   — hyperamylasuria persists for a few days longer than hyperamylasaemia
   — high serum amylase with a normal urine amylase suggests macroamylasaemia
3. Amylase clearance: creatinine clearance ratio
   — increased in acute pancreatitis
   — more specific than serum amylase elevation but sometimes elevated in other conditions
4. Serum trypsin
   — more specific than amylase because it is only produced by the pancreas but it is more difficult to measure (radioimmunoassay).
5. Serum lipase
   — also of diagnostic value, especially in combination with amylase
   — difficult to measure precisely, so amylase tends to be the preferred test (but newer methods may change this)

6. Secondary biochemical abnormalities
   (i) hypocalcaemia—possible mechanisms:
      a. acute illness drops the serum albumin with a consequent reduction in total serum calcium (ionised calcium unchanged)
      b. lipase induced hydrolysis of fat → calcium deposition in tissues by binding with the released fatty acids to form insoluble soaps (fat necrosis)

      N.B. hypercalcaemia which precipitated the pancreatitis may be masked by these changes
   (ii) methaemalbuminaemia (in acute *haemorrhagic* pancreatitis)
   (iii) hyperglycaemia
   (iv) hyperbilirubinaemia

Chronic pancreatic disease

Clinical features:
1. severe epigastric pain (no relief from vomiting or alkalis)
2. weight loss
3. steatorrhoea

Aetiology:
Again poorly understood but may be related to:
1. alcoholism
2. calcification of the pancreas
3. recurrent relapsing acute pancreatitis

Laboratory investigation
1. Secretin test
   — duodenal intubation and collection of duodenal juice to assess the bicarbonate output in response to injection of secretin; cholecystokinin (pancreozymin) may be given as well as secretin to assess enzyme response but this is of doubtful value and increases the side effects.
   — limitations:
      (i) technically difficult
      (ii) may be false positives
      (iii) cannot distinguish between cancer and inflammation (a reduced volume output with normal bicarbonate concentration is said to suggest malignancy)
2. Lundh test
   — measure of tryptic activity in duodenal juice following a test meal
   — again requires duodenal intubation
   — cheaper, simpler and better tolerated than the secretin test
   — limitations:
      (i) again technically difficult

(ii) assumes an intact small intestine
(iii) cannot distinguish between cancer and inflammation
— when performed by an expert:
(i) low trypsin activity is suggestive of extensive pancreatic disease
(ii) normal trypsin activity makes such disease improbable

3. PABA test
   — non-invasive recovery of PABA from urine following chymotrypsin hydrolysis of a synthetic peptide (N-benzoyl-L-tyrosyl-p-aminobenzoic acid) administered orally
   — limitations: false positives may occur in hepatic, intestinal or renal disease.
4. Faecal fat excretion
   — increased in pancreatic malabsorption
   — limitations: the timing of faecal collections is so imprecise that only very high results are significant and this degree of steatorrhoea is usually obvious visually
5. Glucose tolerance test
   — often abnormal in chronic pancreatitis

Radiological aids:
1. ultrasound
2. ERCP

MALABSORPTION

Classification of malabsorption

1. General defects
   (i) Maldigestion
      a. pancreatic insufficiency
         — chronic pancreatic disease
         — cystic fibrosis
      b. deficiency or deconjugation of bile salts
         — bile duct obstruction
         — ileal resection/infiltration (failure of reabsorption of bile salts)
         — small bowel stasis (overgrowth of bacteria which deconjugate bile salts)
   (ii) Malassimilation
      — mucosal damage or loss:
         a. coeliac disease
         b. jejunal resection/infiltration
         c. small bowel stasis
2. Specific defects
   (i) transport system deficiencies, e.g.
      a. glucose/galactose
      b. cystinuria
      c. Hartnup disease

- (ii) enzyme deficiencies, e.g.
  - a. disaccharidase
  - b. enterokinase
- (iii) secretory deficiencies, e.g. intrinsic factor

3. Malabsorption caused by drugs
   - (i) drugs affecting mobility
   - (ii) alcohol (excessive)
   - (iii) neomycin (long term e.g. in liver disease)
   - (iv) cholestyramine (binds bile salts)

Clinical features of malabsorption

1. diarrhoea
2. abdominal discomfort
3. weight loss
4. nutritional deficiencies, e.g.
   - (i) iron
   - (ii) folic acid
   - (iii) vitamin B_{12}
   - (iv) vitamin D
   - (v) vitamin K

Laboratory investigation

Without significant malabsorption biochemical tests are unlikely to be of value and should only used where more reliable tests are unavailable or contraindicated. Histological or radiological evidence is of greater value clinically. Significant malabsorption rarely occurs until there is extensive disease, because of the the considerable functional reserve of the pancreas and intestinal mucosa and because of the adaptation which occurs in non-diseased areas.

1. Pancreatic disease—not a common cause of malabsorption
   - (i) intubation tests (see page 55)
     - — good prediction of extensive disease if expertly performed
   - (ii) PABA text (see page 56)
     - — good screening test for pancreatic disease in patients with steatorrhoea
   - (iii) faecal fat excretion
     - — by the time an unequivocal increase can be demonstrated, the steatorrhoea is usually obvious visually.
2. Jejunal disease
   - (i) jejunal biopsy
     - — really the only satisfactory test
   - (ii) red cell folate
     - — may be used for screening
   - (iii) xylose absorption } very limited value; too many
   - (iv) faecal fat } equivocal results

3. Ileal disease
   (i) bile acid breath test using ^{14}C labelled glycocholic acid
      — measure: $^{14}CO_2$ in expired air
      ^{14}C excretion in faeces
      — good but only available in special centres
   (ii) Schilling test
      — tests absorption of B_{12}
      — readily available
      — abnormal result is also obtained in:
         a. pernicious anaemia
         b. pancreatic disease
         c. bacterial overgrowth
4. Bacterial overgrowth
   — there is no satisfactory biochemical test
   (i) hydrogen breath test } may be normal with positive
   (ii) Schilling test } bacterial cultures

GASTROINTESTINAL HORMONE-SECRETING TUMOURS

These are rare tumours, many of which arise in the pancreas. A single tumour may secrete more than one peptide.

Gastrinoma
— 60% malignant
Gastrinomas are usually islet cell tumours of the pancreas. They cause type I *Zollinger-Ellison* syndrome:
1. high gastric acidity
2. recurrent peptic ulceration (be particularly suspicious of teenagers and the elderly)
3. persistent diarrhoea (some patients)
4. may be part of multiple endocrine adenopathy type I (see page 87)

Type II Zollinger-Ellison syndrome (hyperplasia) of the G cells of the gastric antrum) is even more rare.

Investigation:
Gastric function tests (overnight collection of gastric secretion and the pentagastrin test) have now been largely superseded by the measurement of plasma gastrin.
Plasma gastrin levels may also be high in:
1. hypercalcaemia (which stimulates gastrin release)
2. achlorhydria (lack of inhibitory feedback to gastrin release)
   — atrophic gastritis
   — pernicious anaemia

Insulinoma
— usually benign (see page 67)

Glucagonoma

— 60% malignant

1. characteristic skin rash (skin biopsy shows necrosis and lysis of the epidermis with a normal dermis)
2. mild diabetes
3. 90% occur in women

Vipoma

(Verner-Morrison syndrome)—mostly malignant

These secrete vasoactive intestinal polypeptide (VIP) and are usually islet cell tumours of the pancreas.

1. continous *w*atery *d*iarrhoea
2. *h*ypokalaemia
3. *h*ypochlorhydria

} hence another name: WDHH syndrome

4. dehydration
5. 75% occur in men

Carcinoid syndrome

— mostly benign (see page 86)

Diabetes mellitus and hypoglycaemia

HORMONAL CONTROL OF ENERGY METABOLISM

Insulin

1. Polypeptide hormone (molecular weight about 6000).
2. Secreted by the beta cells of the pancreatic islets.
3. Composed of two chains, A and B, held together by disulphide linkages.
4. Synthesised as a single polypeptide chain (pro-insulin, molecular weight about 9000). A central peptide (C-peptide) is split off to leave the A and B chains of insulin.
5. When insulin is released from the islets, equimolar amounts of C-peptide are released at the same time.

Actions of insulin

1. Reduces hepatic glucose production by:
   (i) reducing glycogenolysis. (Glucagon and adrenaline mobilise glucose from glycogen by activating hepatic phosphorylase. Insulin opposes this)
   (ii) reducing gluconeogenesis—again antagonising glucagon
2. Stimulates glucose transport into cells (except brain, liver and erythrocytes)
3. In adipose tissue, insulin reduces the release of free fatty acids and stimulates the storage of triglyceride. (The consequent reduction in plasma free fatty acids is probably the main reason for the anti-ketotic effect of insulin, though it may also have some direct anti-ketotic effect on the liver)

Other hormones

The other important hormones known to be involved in control of energy metabolism are shown below. They have actions which tend to antagonise insulin.

1. Glucagon
2. Adrenaline
3. Cortisol
4. Growth hormone

DIABETES MELLITUS

Diabetes mellitus is a syndrome characterised by chronically elevated plasma glucose concentration, often accompanied by other clinical and biochemical abnormalities. *Impaired glucose tolerance* is now recognised as an allied but separate condition and there are also various statistical risk categories for diabetes.

Classification of diabetes mellitus and related conditions

1. Diabetes mellitus
   - (i) Insulin-dependent (IDDM) or 'type 1'
     — formerly known as 'juvenile-onset'
   - (ii) Non-insulin-dependent (NIDDM) or 'type 2'
     — formerly known as 'maturity-onset'
     — can be sub-divided into:
       a. obese
       b. non-obese
   - (iii) Secondary diabetes associated with:
     a. Pancreatic diseases
     b. Hormonal disorders (e.g. Cushing's)
     c. Drug or chemical agents
     d. Insulin receptor abnormalities
     e. Genetic syndromes
     f. Miscellaneous
2. Impaired glucose tolerance
3. Statistical risk classes
   - (i) Previous abnormality of glucose tolerance
   - (ii) Potential abnormality of glucose tolerance e.g. relative with diabetes, HLA type with increased risk (see IDDM below).

Insulin-dependent diabetes mellitus

1. Onset usually in childhood or youth
2. Florid symptoms often including weight loss
3. Rapidly progressive course and lethal in the absence of insulin treatment
4. Prone to develop ketoacidosis (see below)
5. Aetiology:
   - (i) genetics—associated with possession of HLA DR3 and DR4
   - (ii) viral infection appears to be the most probable agent linking the genetic susceptibility with beta cell damage and failure perhaps directly or by way of intermediary autoimmune processes (islet cell antibodies)

Non-insulin-dependent diabetes mellitus

1. Onset usually at an older age than IDDM (though there is a distinct subgroup with onset in youth and a strong familial tendency—previously called 'maturity-onset diabetes of youth')

2. Symptoms often less marked than IDDM and often associated with obesity
3. Can often be controlled with diet ± oral hypoglycaemic agents
4. Much less prone to ketoacidosis than IDDM
5. Aetiology:
   (i) genetic susceptibility seems to play an even more important role than in IDDM
   (ii) often associated with tissue resistance to the effects of insulin—obesity often appears to be important in this

Impaired glucose tolerance
1. Mildly impaired glucose tolerance insufficient to warrant a diagnosis of diabetes.
2. Includes those previously called 'chemical', 'borderline', 'subclinical' or 'early' diabetics.
3. Annually, 2–4% of these develop unequivocal diabetes
4. Virtually no risk of clinically significant diabetic retinopathy or nephropathy
5. Double the normal risk of coronary heart disease
6. In pregnancy, some believe it should be treated as diabetes

Metabolic defect in uncontrolled diabetes mellitus
The basic defect is insulin deficiency (relative or absolute). This leads to:
1. Hyperglycaemia caused by:
   (i) increased hepatic glucose production (now believed to be the major cause of hyperglycaemia). This results from unopposed action of glucagon and other hormones on the liver (see 'Actions of insulin' above).
   (ii) impaired peripheral glucose utilisation in turn caused by:
      a. raised circulating free fatty acid levels (see below). Free fatty acids can be used as an alternative substrate by many tissues and when present in sufficient concentration, they inhibit glucose utilisation.
      b. reduced direct action of insulin on glucose uptake in peripheral tissues—this is now thought to be less important than (a).
2. Increased plasma concentration of free fatty acids, ketone bodies and triglycerides (see Fig. 13)
   (i) insulin lack causes lipolysis in adipose tissue thus releasing fatty acids into the circulation
   (ii) increased circulating free fatty acid concentration in turn stimulates hepatic synthesis of ketone bodies and triglycerides

Symptoms of diabetes mellitus

Some or all of the following may be present:

1. Polyuria and polydipsia—if hyperglycaemia is sufficient, the renal tubular maximum for glucose reabsorption will be exceeded leading to glycosuria and a consequent osmotic diuresis. The polydipsia is secondary to this.
2. Weight loss
3. Recurrent infections—related to the consistently high plasma glucose concentrations. Urinary infections related to the glycosuria.
4. Symptoms related to the complications of diabetes (see below)

Diagnostic criteria

1. Random venous plasma glucose of 11 mmol/l or more

or

2. Fasting venous plasma glucose of 8 mmol/l or more

If diabetic symptoms are present, one abnormal value is sufficient to confirm the diagnosis, otherwise any abnormality should be confirmed by repeat testing.

Postprandial values below 8 mmol/l or fasting values below 6 mmol/l on at least two occasions exclude the diagnosis of diabetes.

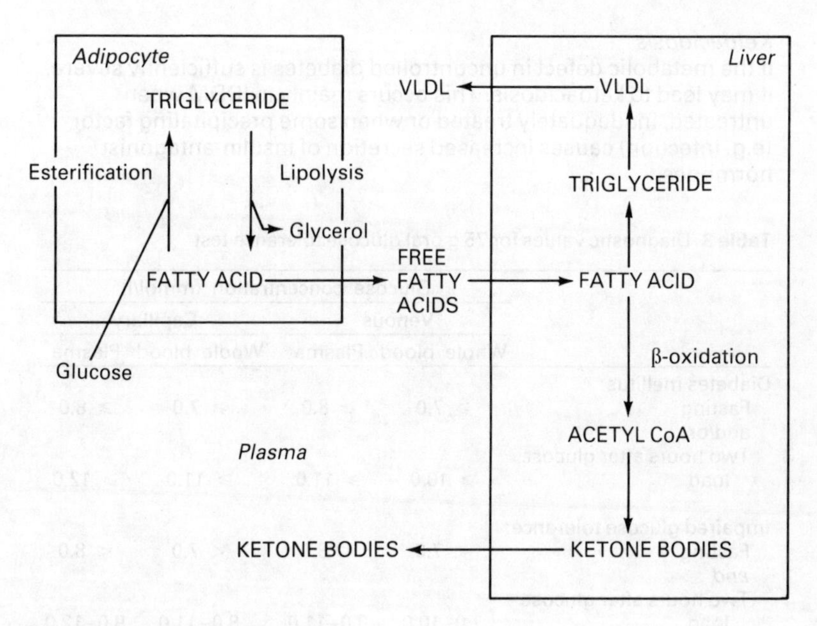

Fig. 13 Ketone body synthesis

An oral glucose tolerance test (see appendix) is indicated:
1. when random or fasting plasma glucose values fall in the equivocal range
2. in pregnancy glycosuria

Table 3 gives diagnostic values for a 75 g oral glucose tolerance test.

Acute complications of diabetes

1. Coma/pre-coma
   (i) Hypoglycaemia
   (ii) Ketoacidosis
   (iii) Hyperosmolar non-ketotic coma
   (iv) Lactic acidosis

   N.B. It should be stressed that non-metabolic comas are just as common in diabetics as non-diabetics and are generally associated with some loss of glycaemic control.
2. Infection
3. Acute neuropathy

Hypoglycaemia

Hypoglycaemia is the most common cause of unconsciousness in the diabetic taking insulin. It may also occur in patients on sulphonylureas, though this is less common.

Ketoacidosis

If the metabolic defect in uncontrolled diabetes is sufficiently severe, it may lead to ketoacidosis. This occurs mainly in IDDM when untreated, inadequately treated or when some precipitating factor (e.g. infection) causes increased secretion of insulin-antagonist hormones.

Table 3 Diagnostic values for 75 g oral glucose tolerance test

	Glucose concentration (mmol/l)			
	Venous		Capillary	
	Whole blood	Plasma	Whole blood	Plasma
Diabetes mellitus:				
Fasting	⩾ 7.0	⩾ 8.0	⩾ 7.0	⩾ 8.0
and/or				
Two hours after glucose load	⩾ 10.0	⩾ 11.0	⩾ 11.0	⩾ 12.0
Impaired glucose tolerance:				
Fasting	< 7.0	< 8.0	< 7.0	< 8.0
and				
Two hours after glucose load	7.0–10.0	8.0–11.0	8.0–11.0	9.0–12.0

Features:

1. Hyperglycaemia—leading to osmotic diuresis which in turn causes:
   (i) sodium and water depletion and hence hypovolaemia
   (ii) potassium depletion—but despite a total body potassium depletion, the plasma potassium may be high before treatment because of a shift of potassium out of cells caused by:
   a. acidosis (see below)
   b. the fact that insulin normally drives potassium into cells
   This shift of potassium out of cells accentuates the urinary potassium loss
   (iii) phosphate depletion
2. Markedly increased ketone body production leading to:
   (i) ketonuria and the smell of acetone on the breath
   (ii) severe *metabolic acidosis*—caused by the ketone bodies: beta-hydroxy-butyrate and acetoacetate. (4+ ketonaemia is necessary to make a diagnosis of ketoacidosis)
   (iii) hyperventilation (Kussmaul respiration)—an attempt to compensate for the metabolic acidosis
3. Impairment of consciousness—related to the metabolic disturbances

Hyperosmolar non-ketotic coma

1. Generally occurs in the elderly
2. Occurs most often in NIDDM
3. Pathogenesis obscure, but many cases are associated with a high carbohydrate intake, e.g. carbohydrate-rich drinks
4. Absence of significant ketoacidosis
5. Hyperglycaemia—(plasma glucose often $>$ 50 mmol/l) → osmotic diuresis → water and electrolyte depletion, water loss often being relatively greater than sodium loss → hypernatraemia (50% have plasma sodium $>$ 150 mmol/l). Dehydration → high plasma urea (often $>$ 20 mmol/l). All these changes → high plasma osmolality (often $>$ 360 mmol/kg)

Lactic acidosis

1. *Rare* cause of metabolic acidosis in the diabetic
2. Usually related to the use of biguanides (phenformin and metformin)—the incidence has dropped sharply with the decreased use of these drugs
3. Plasma glucose may be high, low or normal

Chronic complications of diabetes

1. Diabetic retinopathy
2. Diabetic nephropathy
   — proteinuria is the clinical hallmark
   — once established, progressive deterioration of renal function is inevitable

3. Diabetic neuropathy
   — somatic and autonomic neuropathy
4. Atheroma
   — occurs earlier and more extensively than in the non-diabetic
5. Diabetic foot
   — caused by:
   (i) neuropathic damage
   (ii) ischaemia related to atheroma

Monitoring of treatment

Good control is important in reducing the likelihood of complications.

1. Urine testing
   — the renal threshold for glucose is usually about 10 mmol/l, so urine testing is useless for plasma levels below this
2. Blood/plasma glucose
   — venous blood samples may be taken periodically, but this cannot be performed frequently enough to provide a really reliable assessment of glycaemic control (unless the patient is in hospital)
   — capillary blood samples can be taken by the patient and blood glucose assessed immediately by means of a 'stick test' or dried blood spots (on special filter paper) can be sent to the laboratory through the post
3. Glycated haemoglobin (i.e. glycosylated haemoglobin or haemoglobin A_{1c}
   — during the life-span of the erythrocyte, its contained haemoglobin becomes slowly modified by non-enzymatic reaction with glucose. The rate of this glycation is dependent on the prevailing plasma glucose concentration and once formed the glycated haemoglobin remains in the red cell. The proportion of haemoglobin which is glycated gives an estimate of the glycaemic control over the preceeding 6–8 weeks
   — useful for assessing long term control, but short term swings between hyperglycaemia and hypoglycaemia can only be detected by serial glycose measurements
   — poor index of control in patients with anaemia or increased red cell turnover.

HYPOGLYCAEMIA

Causes

1. Reactive postprandial hypoglycaemia
   (i) Related to gastric surgery
   (ii) Idiopathic
   (iii) Rare causes in infants and children e.g. fructose intolerance, galactosaemia

2. Fasting hypoglycaemia
   (i) Insulinoma
   (ii) Extra-pancreatic tumours especially sarcomas (e.g. retroperitoneal fibrosarcomas)
      — by secretion of a pro-insulin-related peptide
      — tumour usually palpable or visible on X-ray of chest or abdomen
   (iii) Ethanol-induced (Ethanol inhibits gluconeogenesis, but normally blood glucose can be maintained from glycogen reserves. However, if glycogen reserves are low, as in alcoholics who drink but do not eat, hypoglycaemia may occur)
   (iv) Hypopituitarism/adrenal insufficiency
   (v) Severe liver disease
   (vi) Factitious hypoglycaemia
      — patient surreptitiously taking insulin or a sulphonylurea drug
   (vii) Other causes in childhood
      a. temporary neonatal hypoglycaemia as in small-for-dates babies and babies of diabetic mothers
      b. ketotic hypoglycaemia
         — usually in boys aged 1–8 years who often have been small-for-dates when born
         — ketosis (low plasma glucose suppresses plasma insulin, so inducing ketosis)
         — frequent feeding prevents attacks which disappear in adulthood
      c. nesidioblastosis (rare)
         — occurs in the neonate
         — persistent hypoglycaemia
         — no ketosis
         — inappropriately high plasma insulin
         — abnormal 'nests' of beta cells in the pancreas
      d. rare inherited disorders
         — e.g. type I glycogen storage disease, galactosaemia

Diagnosis

Reactive hypoglycaemia:
In suspected reactive hypoglycaemia, an oral glucose tolerance test with samples taken up to 5 hours, will determine if hypoglycaemia develops with coincident symptoms.

Fasting hypoglycaemia:
1. Most causes of fasting hypoglycaemia in adults, other than insulinoma, can be easily recognised clinically.

2. Measurement of plasma insulin during hypoglycaemia is helpful. Inappropriately high levels are found with:
   (i) insulinoma
   (ii) factitious hypoglycaemia
   (iii) nesidioblastosis

   In these situations, the excess insulin prevents ketosis.

Most insulinomas are easily diagnosed by plasma glucose and insulin assays during a presenting hypoglycaemic episode or after an overnight fast. If hypoglycaemia does not occur with an overnight fast (as in 10% of patients), the following are probably now the tests of choice:

(i) induce hypoglycaemia with injected insulin and then measure plasma C-peptide. Normal patients will show suppression of C-peptide, reflecting suppression of their own endogenous insulin release (C-peptide is secreted from the islet cells on an equimolar basis with each insulin molecule) or
(ii) measure fasting pro-insulin (the tumours are always moderately undifferentiated and secrete a high proportion of pro-insulin)—this would be the easiest diagnostic method, but unfortunately, pro-insulin assays are not widely available yet.

Endocrinology

PRINCIPLES OF ENDOCRINE FUNCTION TESTS

Sometimes basal blood or urine samples can be used (e.g. the majority of the currently used thyroid function tests rely on a single basal blood sample).

However in many cases, the traditional approach of stimulation and suppression testing must be followed:

SUSPECTED UNDER-FUNCTION → STIMULATION TEST
SUSPECTED OVER-FUNCTION → SUPPRESSION TEST

Stimulation tests rely on applying a stimulus which normally increases the output of the hormone in question while suppression tests involve applying a stimulus which normally suppresses the hormone. In the latter case, failure of suppression would imply that the secretion of that hormone is not under normal feed-back control.

ANTERIOR PITUITARY

Physiology

The important hormones of the anterior pituitary are:

1. Adrenocorticotrophic hormone (ACTH) or corticotrophin
2. Thyroid stimulating hormone (TSH) or thyrotrophin
3. Growth hormone (GH) (or HGH—human growth hormone)
4. Follicle stimulating hormone (FSH) } the pituitary
5. Luteinising hormone (LH) } gonadotrophins
6. Prolactin

These are all under the control of releasing factors or release-inhibitory factors produced by the hypothalamus. In three cases, the chemical structure of the factors is known and so these factors are called hormones—namely thyrotrophin releasing *hormone* (TRH), luteinising hormone releasing *hormone* (LHRH) and growth hormone release-inhibitory *hormone* (GH-IH).

The hypothalamic factors reach the anterior pituitary via the hypothalamo-hypophyseal portal venous system which passes down the pituitary stalk. Damage to this stalk causes impaired release of all the above pituitary hormones except prolactin, the secretion of which increases.

This is because prolactin is on its own, in being the only one of the above pituitary hormones which is *primarily* under the control of a release-*inhibitory* factor. It seems very likely that this factor is dopamine.

Growth hormone

Growth hormone deficiency
Growth hormone deficiency has no significant effects in adults after growth has ceased, but in children it leads to dwarfism.

GH deficiency may occur as an isolated defect or as part of hypopituitarism (see below).

Tests for growth hormone deficiency:
Basal blood samples are rarely of value—it is the ability of the pituitary to produce GH when stimulated which is in question. The following tests are recommended when *isolated GH deficiency* is suspected:

1. Post-exercise or sleeping GH levels
   — exercise and the early stages of sleep are associated with increased GH secretion
   — a high GH level excludes GH deficiency and no further tests would then be required
2. Clonidine stimulation test (see appendix)
   — a fairly recently introduced test which is safe and reliable
   — clonidine stimulates GH release
3. Insulin stress test (see appendix)
   — the reference test
   — hypoglycaemia stimulates GH release
   — probably only indicated now if the clonidine test shows an abnormally low GH response
   — probably best performed in one of the reference centres for GH deficiency

Growth hormone excess
Cause: GH producing pituitary adenoma

If GH excess occurs during growth, gigantism is the result, while in adults acromegaly occurs.

Tests for growth hormone excess
No difference between testing in gigantism and acromegaly.

1. Basal resting GH level
   — a low level virtually excludes the diagnosis
2. Oral glucose tolerance test with measurement of GH on all samples
   — an elevated plasma GH which fails to suppress during the test confirms the diagnosis

Other biochemical features of GH excess
Some patients show:
1. *Hypopituitarism*
   — acromegaly or gigantism may be associated with deficiency of the other anterior pituitary hormones (caused by compression as a result of the GH producing adenoma). Therefore, if GH excess is confirmed, a *triple pituitary stimulation test* should always be performed (see 'Hypopituitarism').
2. Abnormal glucose tolerance (one-quarter of patients)
3. Elevated serum phosphate

Hypopituitarism

Causes
1. Pituitary tumours—pituitary adenomas, craniopharyngiomas
2. Malignant disease—secondary carcinoma, local cerebral tumours
3. Infectious diseases—e.g. TB meningitis
4. Granulomatous diseases—e.g. sarcoidosis
5. Vascular disease—e.g. post-partum necrosis (Sheehan's syndrome)
6. Iatrogenic—hypophysectomy
7. Trauma—head injury
8. Secondary to hypothalamic disease
   e.g. craniopharyngiomas and other tumours
   — usually associated with diabetes insipidus
   — rarely associated with pathological hunger and obesity or loss of appetite and severe malnutrition
9. 'Functional' hypothalamic disorders
   — secondary to malnutrition (e.g. anorexia nervosa)
   — usually a reversible impairment of pituitary gonadotrophin production only

Sequence of anterior pituitary hormone loss
The ability to secrete the anterior pituitary hormones is normally lost in the following sequence (certainly when caused by a space-occupying lesion in the pituitary):
1. Gonadotrophins and GH
2. ACTH and TSH

Thus one of the early features is reproductive dysfunction (amenorrhoea, infertility, impotence). In post-menopausal women, normal postmenopausal levels of LH and FSH (i.e. high levels) are very much against a diagnosis of hypopituitarism.

Tests for hypopituitarism
Tests for hypopituitarism rely largely on stimulation tests (see 'Principles of Endocrine Function Tests'). Table 4 lists the stimuli for the release of the various anterior pituitary hormones.

1. Insulin stress test* (see appendix)
   — tests for GH and ACTH deficiency
   — ACTH response generally assessed via the cortisol response
   — prolactin release is also stimulated, but this is not normally measured
2. TRH test* (see appendix)
   — tests TSH release
3. LHRH test* (see appendix)
   — tests LH release
   — FSH is also released, but not normally measured
4. Basal thyroid function tests
5. Basal prolactin
   — high with prolactinomas (see below) or when a disease prevents the hypothalamic factors from reaching the pituitary (see 'Anterior Pituitary—Physiology').

* these three may be combined into a single test—the *triple pituitary stimulation test* (see appendix)

Prolactinomas and hyperprolactinaemia

Many of the pituitary tumours which were previously thought to be non-functioning have now been shown to be producing prolactin.

Features of prolactinomas

1. Reproductive dysfunction
   — infertility, amenorrhoea in women
   — impotence in men

Table 4 Important stimuli causing release of anterior pituitary hormones

Hormone	Stimulus
ACTH	— Corticotrophin releasing factor — Low serum cortisol — Stress (including *hypoglycaemia*) — Pyrogens
TSH	— *TRH* — Low serum T4 & T3
GH	— GH releasing factor — Stress (including *hypoglycaemia*)
Gonadotrophins	— *LHRH* (LH releasing hormone) — Clomiphene
Prolactin	— Lack of prolactin release inhibitory factor (probably dopamine) — Stress — TRH — Anti-dopaminergic drugs (e.g. phenothiazines)

— caused by a resistance of the gonads to the effects of gonadotrophins (induced by high serum prolactin levels) perhaps combined with hypopituitarism
2. Galactorrhoea
3. Features related to the pituitary tumour itself
   — visual field defects, headaches etc.
4. Hypopituitarism (see above)
   — caused by compression of the normal pituitary

Causes of hyperprolactinaemia
1. Stress
2. Drugs e.g. phenothiazines
3. Hypothalamic and pituitary stalk disease
   — by preventing prolactin release inhibitory factor from reaching the anterior pituitary
4. Pituitary tumours
   — prolactinomas
   — other pituitary tumours (presumably by the same mechanism as in 3. above)
5. Primary hypothyroidism
   — TRH also releases prolactin
6. 'Ectopic' prolactin secretion (by a tumour of non-endocrine tissue)
7. Chest wall injury
   — presumably by stimulating the afferent nerve pathway normally involved in the stimulation of prolactin release during suckling

THYROID

Physiology

The hormones secreted by the thyroid gland are:
1. Thyroxine (T4)
   0.02% free in serum, the rest is protein bound to:
   — thyroxine binding globulin (TBG)
   — thyroxine binding prealbumin
   — albumin
2. Tri-iodothyronine (T3)
   0.36% free in serum

N.B. It is the free hormone which is biologically active
In serum: [free T4] = 4 × [free T3]
Metabolic activity of T3 is approximately 4 × that of T4
Thus biological activity of T4 and T3 are about equal

Secretion of T4 and T3 is stimulated by Thyroid Stimulating Hormone (TSH or thyrotrophin) which is released by the anterior pituitary under control of Thyrotrophin Releasing Hormone (TRH)—see 'Anterior Pituitary'

The *free* thyroid hormones in turn suppress the release of TRH and TSH, thus completing a feed-back control loop.

N.B. Changes in thyroid hormone binding proteins in serum will result in comparable changes in *total* T4 and *total* T3 concentrations because the above feed-back loop will keep the *free* hormone concentrations steady.

Though some T3 is secreted directly by the thyroid, much is produced by peripheral conversion of T4 and T3. Some drugs (e.g. beta-blockers and amiodarone) impair this peripheral conversion and may lead to confusing thyroid function test results.

Actions of thyroid hormones
Numerous actions on the metabolism of carbohydrates, proteins and fats such that:

1. they give rise to an increased basal metabolic rate
2. they are essential for normal growth and development
3. they seem to increase the sensitivity of beta-receptors to catecholamines

Thyroid function tests

1. Serum total T4 and total T3
   — include protein-bound and free hormone
   — levels affected by changes in binding proteins
   — generally measured by immunoassay, so no longer subject to interference from iodine contamination
2. T3 uptake
   — Causes endless confusion!
   — Nothing to do with serum T3 concentration!
   — Radioactively labelled T3 is used as a laboratory reagent to assess the level of unoccupied binding sites on the thyroid hormone binding proteins. Labelled T3 not taken up by these binding sites is removed by a resin. Radioactivity is then counted in either serum or resin fractions.
   — T3 uptake may be expressed as either T3 *serum* uptake or T3 *resin* uptake. T3 serum uptake is *directly* proportional to unoccupied binding sites while T3 resin uptake is *inversely* proportional to unoccupied binding sites. Make sure you know which version your laboratory uses.
3. Free thyroxine index (FTI)
   — the result of dividing the total T4 by the T3 serum uptake (or multiplying the total T4 by the T3 resin uptake)
   — it is a result without units which is intended to be proportional to the free T4 concentration
   — unfortunately it is not entirely reliable but does give some correction for changes in total T4 which result from changes in binding proteins.

4. Serum free T4 and free T3
   — direct measurement of free hormone concentrations in serum
   — now fairly widely available (especially free T4)
   — generally not affected by changes in binding proteins, though some assays are mildly affected by changes in serum albumin concentration
5. Serum TSH
   — the assays in general use are unable to differentiate low levels from normal and so are only of real value in the diagnosis of hypothyroidism (see below)
   — newer more sensitive assays being introduced are capable of this differentiation and are of use in the diagnosis of hyperthyroidism as well
6. TRH test (see appendix)
   — a dynamic function test that involves measuring the response of serum TSH levels to an i.v. injection of TRH

Use of thyroid function tests
The policy of laboratories varies but might be something like:
1. either
   (i) Measure total T4
   (ii) If there is any suspicion of binding protein abnormalities measure T3 uptake and calculate FTI

   or

   Measure free T4
2. If borderline results are obtained in 1. or there is a strong clinical suspicion of thyroid disorder
   (i) In suspected hypothyroidism measure TSH*
   (ii) In suspected hyperthyroidism measure total or free T3*

   * see the appropriate section below for an explanation
3. If there is still doubt, perform a TRH test (mainly of value in suspected hyperthyroidism)

N.B. The new 'low level' TSH assays may soon change the whole approach to thyroid function testing (perhaps becoming the first-line test)

Hypothyroidism

Causes
1. Primary hypothyroidism
   (i) Autoimmune thyroiditis
   (ii) Damage to the thyroid
      a. Surgery
      b. Radiation (after ^{131}I treatment of hyperthyroidism)
   (iii) Rare causes—e.g. goitrogens, endemic cretinism (related to iodine deficiency), dyshormonogenesis (inherited disorders of thyroid hormone synthesis)

2. Secondary hypothyroidism
   Pituitary or hypothalamic disease impairing the release of TRH or TSH

Laboratory diagnosis

1. Total T4
   — total T4 usually low or low normal
   — T3 is less sensitive than T4 for detecting hypothyroidism, so its measurement is not useful here
2. T3 uptake
   — T3 *serum* uptake usually high (T3 *resin* uptake usually low) indicating more unoccupied binding sites
3. Free thyroxine index
   — usually low
4. Free T4
   — free T4 low or low normal
   — similar to total hormone measurements, free T3 measurement is not useful
5. TSH
   — high in primary hypothyroidism
6. TRH test
   — increased TSH response in primary hypothyroidism but the basal TSH is usually high, so this test is rarely needed in this situation
   — usually reduced or delayed response in secondary hypothyroidism

Other laboratory findings

Many patients will show:

1. Raised serum cholesterol
2. Raised serum creatine kinase and sometimes a raised aspartate aminotransferase—released from muscles
3. Macrocytosis
4. Raised serum prolactin (in primary hypothyroidism)
   — increased TRH production by the hypothalamus not only releases TSH from the anterior pituitary, but prolactin as well (see 'Anterior Pituitary')

Treatment and its monitoring

— Primary hypothyroidism

Thyroxine tablets—normally 0.1 to 0.2 mg/day

The half-life of serum thyroxine is approx. 7 days, so there is no point in dividing the dose; thyroxine should always be given as a once daily dose. With all drugs it takes 3 to 5 half lives for a steady state to be achieved, so one should wait at least 4 weeks after starting thyroxine before performing any tests to assess the adequacy of replacement. After that time, TSH should be measured to check that it has been suppressed to normal.

After any dosage reduction, if test are to be performed to check that the new replacement dose is adequate, at least 6 weeks should be allowed to elapse and preferably longer. This is because it may take that long for the ability to secrete TSH to return.

Tri-iodothyronine is usually preferred for the initial treatment of myxoedema coma because of a quicker onset of action and a shorter half-life should the dose need to be reduced.

— *Secondary hypothyroidism*

Similar to primary hypothyroidism but *thyroxine must never be started until hypoadrenalism has been treated or excluded* because of the risk of precipitating adrenal crisis. Also TSH measurement is clearly of no value in assessing replacement and one must rely on the serum T4 or T3 level(s) alone—a good clinical assessment is probably more valuable.

Hyperthyroidism

Causes

1. Graves' disease
   — the TSH receptors of the thyroid appear to be stimulated by an autoimmune process
   — most cases seem to be caused by circulating 'thyroid stimulating immunoglobulins' (TSI)
2. Toxic multinodular goitre
3. Toxic adenoma
4. Over-treatment with thyroxine
5. Very rare causes—e.g. ectopic TSH syndrome, struma ovarii (ovarian teratoma containing mainly thyroid tissue).

Laboratory diagnosis

1. Total T4 and total T3
   — total T4 is usually high or high normal
   — T3 is *more* sensitive than T4 for detecting hyperthyroidism (cf hypothyroidism)
2. T3 uptake
   — T3 serum uptake usually low (T3 resin uptake usually high) indicating fewer unoccupied binding sites
3. Free thyroxine index
   — usually high
4. Free T4 and free T3
   — free T4 usually high
   — free T3 almost always high
5. TSH
   — assays in general use are unable to differentiate low TSH from normal and so basal TSH is useless in the diagnosis of

hyperthyroidism except in the very rare cases which are TSH driven
— in the future, more sensitive assays may be very useful
6. TRH test
— a diminished TSH response to TRH is always present

N.B. T3 thyrotoxicosis—In some patients with hyperthyroidism, the only finding in a basal serum sample is a raised T3. If left untreated these patients will generally develop the other biochemical features hyperthyroidism. Many believe that this is the natural history of hyperthyroidism, though the majority of patients do not present at this early stage.

Graves' disease and pregnancy
Neonatal thyrotoxicosis may occur in the baby born to a mother suffering from Graves' disease if she has high levels of TSI. Like other IgG molecules, TSI are transferred across the placenta to the fetus and the neonate will then have thyrotoxicosis until these TSI disappear.

ADRENAL CORTEX

Physiology

The adrenal cortex produces three groups of hormones:

1. Glucocorticoids—principally cortisol
   — effects on carbohydrate, protein and fat metabolism (in particular they increase gluconeogenesis).
   — immunosuppressive and anti-inflammatory effects.
2. Mineralocorticoids—principally aldosterone
   — increase sodium reabsorption in the distal renal tubule in exchange for potassium and hydrogen ion
3. Androgens—dehydroepiandrosterone (DHA), androsterone and testosterone

Fig. 14 outlines their synthetic pathways.

Cortisol

1. Cortisol has a diurnal variation, levels being highest at about 9 am and lowest at midnight
2. About 95% of plasma cortisol is protein bound, mainly to transcortin
3. Corticotrophin releasing factor (CRF) from the hypothalamus stimulates ACTH release from the anterior pituitary (see 'Anterior Pituitary') and this in turn stimulates cortisol release from the adrenal. Free cortisol (i.e. non-protein bound) causes negative feed-back to pituitary and hypothalamus, thus closing a control loop.

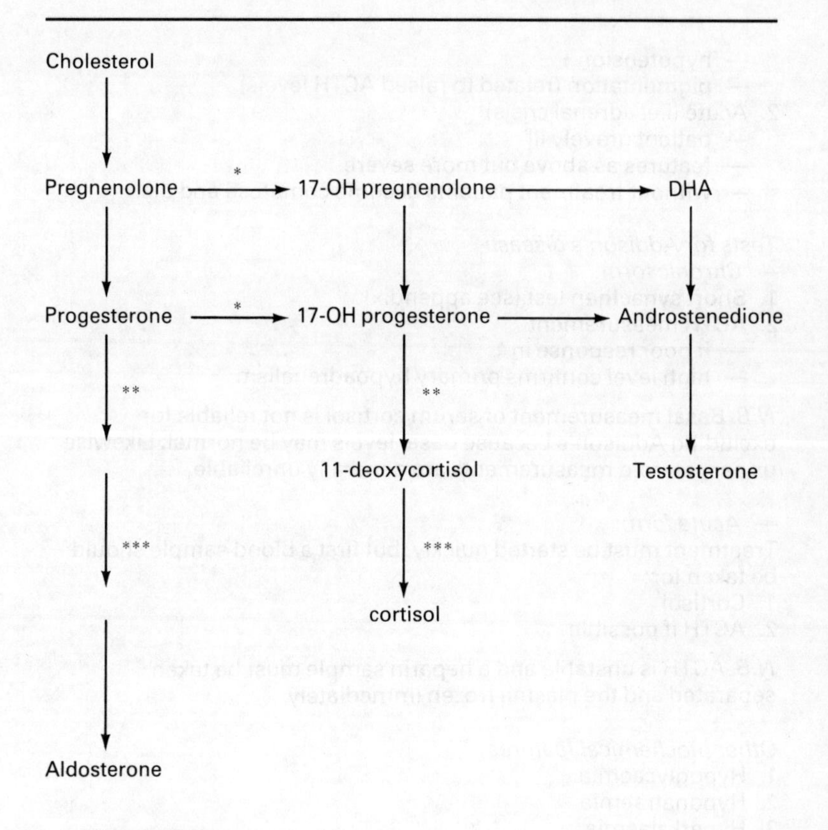

Fig. 14 Biosynthesis of adrenal steroids

Hypoadrenalism

Causes

1. Primary (Addison's disease)
   — autoimmune adrenalitis
   — tuberculous disease of the adrenal
2. Secondary (see 'Hypopituitarism')

Features of Addison's disease

1. Chronic
   — tiredness
   — G.I. complaints

— hypotension
— pigmentation (related to raised ACTH levels)
2. Acute (i.e. adrenal crisis)
— patient gravely ill
— features as above but more severe
— without treatment patients become comatose and die

Tests for Addison's disease
— *Chronic form*:
1. Short synacthen test (see appendix)
2. ACTH measurement
— if poor response in 1.
— high level confirms primary hypoadrenalism

N.B. Basal measurement of serum cortisol is not reliable for excluding Addison's because basal levels may be normal. Likewise urinary steroid measurements are generally unreliable.

— *Acute form*:
Treatment must be started quickly, but first a blood sample should be taken for:
1. Cortisol
2. ACTH if possible

N.B. ACTH is unstable and a heparin sample must be taken, separated and the plasma frozen immediately

Other biochemical features
1. Hypoglycaemia
2. Hyponatraemia
3. Hyperkalaemia
4. Raised blood urea (reduced GFR because of hypotension)
5. Hypercalcaemia

N.B. these features are often *not* present except in the more severe cases.

Cushing's syndrome

Causes
1. Cushing's *disease* (i.e. the pituitary dependent form of Cushing's *syndrome*)
— acts as if there were a resetting of the pituitary feed-back control mechanisms to maintain a higher level of cortisol than normal.
2. Adrenal tumours
— adenoma
— carcinoma
3. Ectopic ACTH (ACTH production by a tumour)

Tests for Cushing's syndrome

— *Diagnosis*:

1. Midnight and 9am serum cortisol levels
   — the normal diurnal variation is usually lost and may even be reversed—thus the midnight level is high.
   — stress may easily elevate the cortisol levels and the midnight level should never be taken on the patient's first night in hospital.
   — usually better to proceed straight to one of the following tests which can be done without admitting the patient to hospital.
2. Overnight dexamethasone suppression test
   — give 1.5 mg of dexamethasone by mouth at 11pm
   — take a blood sample for cortisol at 9am the next morning
   — in normal patients the cortisol level will be similar to normal midnight levels while in Cushing's syndrome it fails to suppress
   — in doubtful cases repeat the 9am cortisol measurement after giving dexamethasone 0.5 mg qds for 2 days (i.e. the standard low dose dexamethasone suppression test)
3. 24 hour urinary free cortisol measurement
4. Insulin stress test (with measurement of serum cortisol)
   — not normally required unless the patient has depression (see below)

— *Differentiation of causes of Cushing's syndrome*:

1. ACTH measurement at 9am (N.B. ACTH is unstable—special sampling conditions are required)
   — suppressed to undetectable levels with adrenal tumours
   — *usually* markedly elevated with ectopic ACTH
   — levels are between the mid point of the reference interval and twice the upper limit with pituitary dependent Cushing's
2. High dose dexamethasone suppression test
   — 2 mg qds for 3 days
   — classically cortisol levels suppress with pituitary dependent Cushing's but not with the other causes (but ectopic ACTH sometimes suppresses)
3. Metyrapone test
   — in most cases, little extra information is obtained from this test but it is mentioned for completeness.

Other biochemical features

1. Sodium retention
   — plasma sodium usually high normal
   — hypertension
2. Hypokalaemia } often but not always
3. Glucose intolerance }

Depression
Endogenous depression may give rise to many of the biochemical features of Cushing's syndrome, even failure of suppression of serum cortisol with dexamethasone. The only test which reliably differentiates depression from Cushing's syndrome is the insulin stress test. Normal and depressed patients show a rise in cortisol with hypoglycaemia, patients with Cushing's do not.

Hyperaldosteronism
1. Primary (i.e. Conn's syndrome)
   — aldosterone producing tumour of the adrenal or adrenal hyperplasia with increased aldosterone production (mechanism unclear)
   — *renin suppressed*
   — features: hypertension with a tendency to hypokalaemia and alkalosis.
2. Secondary
   — generally caused by *increased renin production*
   (i) Causes with a tendency to reduced blood volume
      a. sodium losing states
      b. oedema states (see page 9)
   (ii) Impaired renal blood supply
      — features: as primary hyperaldosteronism
      a. renal artery stenosis
      b. malignant hypertension (hyperplastic arteriolosclerosis and fibrinoid necrosis of the renal arterioles)

N.B. Hyperkalaemia also stimulates aldosterone release but this is usually obvious.

Tests for Conn's syndrome
Consider alternative causes for hypertension and hypokalaemia first (see 'Hypertension').

Most anti-hypertensive drugs interfere and make interpretation of tests very difficult, so they should be stopped, if possible 3 weeks before the tests. Any potassium depletion should be made good by giving potassium supplements, but these should then be stopped before the tests are performed. The patient should be on a normal potassium and sodium intake during the test period. The protocol for testing should be discussed with the laboratory but will probably include some or all of the following:
1. 24 hour urine for aldosterone, sodium and potassium excretion
2. Recumbent and ambulant blood samples for renin estimation
3. Recumbent and ambulant blood samples for aldosterone estimation

N.B. Renin and aldosterone secretion are posture dependent, being lower when lying than when standing.

Congenital adrenal hyperplasia (adreno-genital syndrome)
This is a group of disorders in which there is a deficiency of one of the enzymes on the cortisol biosynthetic pathway. This results in impaired cortisol production and hence increased ACTH production, often sufficient to produce fairly normal levels of cortisol, but at the expense of marked build up of precursors on the pathway before the block. By far the most common is a deficiency of the 21-hydroxylase and only this will be described here.

21-hydroxylase deficiency
Referring to Fig. 14, it will be obvious that 17-OH progesterone will build up in this condition. Since 17-OH progesterone is a precursor of the androgens androstenedione and testosterone, these are also produced in excess. Female infants are masculinised and male infants may show precocious sexual development.

Severe forms of this condition may present in infancy with a salt-losing state because of impaired aldosterone synthesis.

In females, mild forms occasionally do not present until shortly after puberty, when hirsutism and virilisation are the presenting features.

— *Tests*:
1. Plasma 17-OH progesterone
2. 24 hour urine for pregnanetriol
   — pregnanetriol is the major urinary metabolite of 17-OH progesterone
   — this is the preferred test for the mild forms presenting later in life

— *Treatment*:
Treatment is by giving sufficient steroid replacement to suppress the excess ACTH production and hence the excess androgens. However, if too much is given, it will suppress growth. It is currently believed that measurement of plasma *androstenedione* is the best way of assessing adequacy of replacement.

ENDOCRINOLOGY OF THE REPRODUCTIVE SYSTEM

Amenorrhoea
1. Primary amenorrhoea (i.e. the patient has never had periods)
   (i) congenital abnormality of the reproductive tract
      — this includes *testicular feminisation* in which the patient has a male chromosome pattern and is biochemically like a normal male. There seems to be end organ resistance to testosterone, so the patient's external appearance is like that of a normal female.

(ii) Turner's syndrome (chromosome pattern: XO)
— normal ovaries are absent
— low oestrogens/high gonadotrophins

Any of the causes of secondary amenorrhoea may also cause primary amenorrhoea.

2. Secondary amenorrhoea (i.e. the patient has had periods but they have stopped)
   (i) pregnancy
   (ii) primary ovarian failure
   — low oestrogens/high gonadotrophins
      a. normal menopause
      b. autoimmune damage to the ovary
   (iii) hyperprolactinaemia (see page 73)
   (iv) hypopituitarism (see page 71)

Female infertility
This is a very complex subject which can only be covered in outline.

If the patient has amenorrhoea investigations should be directed towards the elucidation of its cause (see above). Otherwise, it can be helpful to divide the causes of infertility into those with normal ovulation and those without. An extremely useful guide to whether the patient is ovulating can be obtained by measuring the serum progesterone level at about day 21 of the menstrual cycle—corpus luteum formation is associated with high levels.

1. With normal ovulation—biochemistry of little further help
2. Without normal ovulation
   (i) measure gonadotrophins
   — high in primary ovarian failure
   (ii) measure prolactin
   — see 'Hyperprolactinaemia'
   (iii) if there is hirsutism or evidence of virilisation perform the relevant investigations (see below).
   (iv) if there is suspicion of hypopituitarism perform the relevant investigations (see page 72)

Hirsutism

Causes

— *Common*:

1. Idiopathic and racial
   — never any other evidence of virilisation
2. Polycystic ovary syndrome

— *Rare*:

3. Cushing's syndrome (see page 80)
4. Mild forms of congenital adrenal hyperplasia (see page 83)
5. Androgen producing tumours

Tests for hirsutism
If there is no other evidence of virilisation (i.e. periods are normal, there is no enlargement of the clitoris etc.) and the patient is of a race more prone to hirsutism or there is a family history of hirsutism, no further investigation is necessary.

If there is clinical evidence of Cushing's syndrome perform the investigations for that.

Otherwise:

1. Plasma testosterone estimation
2. Plasma luteinising hormone (LH)—high in polycystic ovary syndrome (mechanism not clear)
3. 24 hour urine for:
   (i) pregnanetriol (to exclude congenital adrenal hyperplasia)
   (ii) 17-oxo-steroids
   — 17-oxo-steroids in the urine are steroids which represent androgen metabolites.
   — alternatively measure serum DHA-sulphate if available

High 17-oxo-steroid excretion (or serum DHA-sulphate) or significantly elevated plasma testosterone should lead to investigations to exclude an androgen producing tumour. This is a complex situation and further advice should be obtained. Investigation will probably involve assessing the suppression of plasma testosterone and urinary 17-oxo-steroids by dexamethasone 0.5 mg qds given for at least 3 days.

High LH should be followed by a gynaecological consultation with a view to confirming the polycystic ovaries, probably by laparoscopy.

POSTERIOR PITUITARY

The hormones of the posterior pituitary include:

1. Antidiuretic hormone (ADH) also called arginine vasopressin
2. Oxytocin

They are produced by neuronal cells of the para-ventricular and supra-optic nuclei of the hypothalamus. From here they pass down the axons of these cells which terminate in the posterior pituitary. Transection of the pituitary stalk produces a temporary deficiency of these hormones, but the hypothalamic cells regain the ability to release them usually within a few days.

Antidiuretic hormone

The physiology of antidiuretic hormone is covered in chapter II.

Deficiency leads to *diabetes insipidus* in which polyuria is the main presenting feature. The patient may pass 5 to 20 or more litres of urine per day.

N.B. Another form of diabetes insipidus exists in which the kidney is insensitive to the action of ADH (so called nephrogenic diabetes insipidus to differentiate it from this cranial form). See page 15 for the differential diagnosis of polyuria.

Causes of cranial diabetes insipidus:
1. Damage to the hypothalamus or pituitary (e.g. by tumours, surgery in this region, fractures at the base of the skull)
2. Pituitary stalk transection (usually from head injuries) — temporary (see above)

Excess ADH leads to the syndrome of inappropriate ADH secretion (see page 7).

CARCINOID SYNDROME

Caused by tumours which produce 5-hydroxytryptamine (5HT) and related compounds. The most common site for such tumours is the ileo-caecal region, but with such tumours, the carcinoid syndrome only appears after metastasis has occurred (usually to the liver), probably because before metastasis, the liver inactivates the tumour products. With a tumour in other sites (e.g. bronchus), the syndrome appears earlier.

Features of carcinoid syndrome
1. Flushing (paroxysmal)
2. Watery diarrhoea
3. Wheezing

Tests for carcinoid syndrome
Most commonly, the urine is examined for excess 5-hydroxyindole-acetic acid (5HIAA), which is a metabolic product of 5HT.

ADRENAL MEDULLA

The adrenal medulla produces catecholamines, those of physiological importance being:
1. Adrenaline
2. Noradrenaline
3. Dopamine

Phaeochromocytoma
This is a tumour of the adrenal medulla which produces catecholamines.

Features:
1. Hypertension
   — often paroxysmal
   — often presenting at a relatively young age (i.e. 20–50 years)
2. Attacks of headache, sweating, tachycardia, anxiety, nausea and pallor

N.B. There is an association between *neurofibromatosis* and phaeochromocytoma, so patients with neurofibromatosis and hypertension should receive careful investigation to exclude this condition.

Tests:
Generally 24 hour urine(s) for:
1. Hydroxymethoxymandelic acid (HMMA)*
   — often incorrectly called vanilmandelic acid (VMA)
2. Metanephrines*
3. Catecholamines

* metabolites of adrenaline and noradrenaline

N.B. If there is a strong clinical suspicion at least two of the above should be estimated, preferably on more than one 24 hour urine.

Neuroblastoma
1. Common tumours in children, 60% occurring in the first 3 years of life.
2. They generally produce excess noradrenaline and dopamine
3. Hypertension occurs in many cases

Tests:
As for phaeochromocytoma, though because neuroblastomas frequently produce excess dopamine, the dopamine metabolite homovanillic acid (HVA) should also be measured in the urine if possible.

MULTIPLE ENDOCRINE ADENOPATHY (MEA) (or Pluriglandular Syndrome)

In these rare syndromes, two or more endocrine glands secrete inappropriate amounts of hormone, usually from adenomas. The tendency to develop the syndrome is inherited, families developing either MEA I or MEA II:

MEA I
This involves the following (in decreasing order of incidence):
1. Parathyroid gland (hyperplasia or adenoma)
2. Pancreatic islet cells
   (i) gastrinomas
   (ii) insulinomas

3. Anterior pituitary gland
4. Adrenal cortex
5. Thyroid

MEA II
1. Medullary carcinoma of the thyroid
2. Phaeochromocytoma
3. Parathyroid adenoma or carcinoma

Calcium and metabolic bone disease

PHYSIOLOGY

Plasma/serum calcium

1. Approx. 50% is ionised
2. Most of the remainder is protein-bound, especially to albumin
3. Total calcium should always be corrected for albumin changes:
   — correct to an albumin level in the middle of the reference interval (normally 40 g/l)
   — 1 g/l of albumin corresponds to 0.02 mmol/l of calcium
   — for low albumin level : add the adjustment to the measured total calcium
   — for high albumin level : subtract the adjustment

   e.g. Total calcium = 2.0 mmol/l
   Albumin = 30 g/l

   Corrected calcium = 2.0 + ((40 − 30) × 0.02) mmol/l
   = 2.2 mmol/l

N.B. Such corrections should only be considered approximations; the lower the serum albumin, the greater is the imprecision.

Parathyroid hormone (PTH)

1. Produced by the parathyroid glands in response to a reduction in plasma ionised calcium
2. Actions:
   (i) reduces renal tubular reabsorption of phosphate → increased urinary phosphate loss and increased hydroxylation of vitamin D (see below)
   (ii) increases renal tubular reabsorption of calcium
   (iii) stimulates osteoclastic bone resorption

Vitamin D

1. Two sources:
   (i) diet
   (ii) skin (from ultra-violet light exposure)
2. Hydroxylated in the liver to 25-hydroxycholecalciferol (25-HCC)

3. Hydroxylated in the kidney to 1,25-dihydroxycholecalciferol (25-HCC)
   — biologically active form of the vitamin
   — the activity of renal 1-alpha-hydroxylase (the enzyme involved) seems to be increased by low renal intracellular phosphate levels
   — PTH increases the production of 1,25-DHCC probably by decreasing the renal intra-cellular phosphate level.
4. Main action of 1,25-DHCC: increases intestinal absorption of calcium

Control of plasma calcium level
1. Decrease in plasma ionised calcium stimulates PTH production
2. PTH returns the calcium level to normal
   — under physiological conditions, the main reason for this is probably its stimulation of 1,25-DHCC production and consequent increase in intestinal absorption of calcium. (The effect of PTH on bone resorption is probably only important in hyperparathyroidism)
3. Increase in plasma ionised calcium inhibits PTH release, thus lowering the calcium level
   (*N.B.* the secretion of calcitonin, a hormone produced by the C-cells of the thyroid, is increased by hypercalcaemia, but it seems to be of little importance in human physiology)

HYPERCALCAEMIA

N.B. Further discussion assumes that the hypercalcaemia is still present after correction for serum albumin level (see 'Physiology' above)

Symptoms
1. Polyuria and thirst
   — caused by defective renal tubular concentrating ability
2. Weakness and lassitude
3. Confusion and inability to concentrate
4. Constipation
5. Anorexia and vomiting

Causes
1. Malignancy
   (i) bone metastases
   (ii) multiple myeloma
   (iii) humoral hypercalcaemia of malignancy
      — i.e hypercalcaemia in patients with cancer but no evidence of bone metastases
      — probably several humoral factors involved

2. Hyperparathyroidism (see below)
3. Rarer miscellaneous causes:
   (i) Sarcoidosis
   (ii) Hyperthyroidism
   (iii) Vitamin D intoxication
   (iv) Immobilisation (especially with Paget's disease)
   (v) Milk-alkali syndrome
   (vi) Idiopathic hypercalcaemia of infancy
   (vii) Familial hypocalciuric hypercalcaemia (see below)
   (viii) Addison's disease
   (ix) Acromegaly

Differentiation of the causes of hypercalcaemia

1. History and clinical examination are very important
2. Look for evidence of malignancy—biochemical tests helpful here are:
   (i) *serum protein electrophoresis*
      — paraprotein → probable multiple myeloma (see page 135)
      — increased alpha-1 and alpha-2 globulins → acute phase reaction (often present with malignancy)
      — decreased gamma globulin → immuno suppression (suggests malignancy)
   (ii) urine for *Bence-Jones protein* (see 'multiple myeloma' page 135)
   (iii) *plasma PTH* suppressed to undetectable levels
   (iv) high gamma-glutamyl transferase (often the first biochemical evidence of hepatic metastases)
   (v) high acid phosphatase → prostatic malignancy
3. Look for evidence of hyperparathyroidism:
   (i) perform the tests for hyperparathyroidism listed below
   (ii) the following are also pointers to hyperparathyroidism which may be useful while awaiting the results of PTH assay or steroid suppression:
      a. low serum phosphate
      b. hyperchloraemia acidosis (PTH has an effect on hydrogen ion transport by the renal tubule)
4. Perform tests for hyperthyroidism (see page 77), Addison's disease (see page 80) or acromegaly (see page 70) if any of these seem likely clinically.

Hyperparathyroidism

1. Primary
   (i) parathyroid adenoma/carcinoma } → autonomous PTH production
   (ii) parathyroid hyperplasia }
2. Secondary
   — occurs in any condition which tends to lead to chronic *hypocalcaemia* e.g. vitamin D deficiency, renal failure

— unlike primary and tertiary hyperparathyroidism the serum calcium is low or low normal

3. Tertiary
   — an occasional complication of long-standing secondary hyperparathyroidism in which the prolonged stimulation eventually leads to autonomous PTH production

Primary hyperparathyroidism

— *Presentations*:

1. Renal calcium stone disease (see page 31)
   — a common presentation
2. Hypercalcaemia
   — symptoms as above but:
   — often asymptomatic — found on biochemical screening
   — an increasingly common presentation
3. Metabolic bone disease
   — a rare presentation
   — presents as cystic swelling of a bone or a fracture

— *Tests for hyperparathyroidism*:

1. Serum calcium
   — almost always elevated, especially if measured more than once
   — very rarely patients may be normocalcaemia if there is combined vitamin D deficiency
2. Plasma PTH
   — serum calcium must be measured simultaneously
   — level inappropriate to the serum calcium in hyperparathyroidism
   N.B. any measurable PTH is inappropriate if the serum calcium, *corrected for albumin*, was definitely high at the time of measurement
   — if the PTH is consistent with hyperparathyroidism, consider familial hypocalciuric hypercalcaemia (see below)
3. Steroid suppression test
   — classically, in hyperparathyroidism, hypercalcaemia will not suppress when steroids are given for 10 days (usually hydrocortisone 40 mg tds)
   — with most other causes of hypercalcaemia, there is a pronounced fall towards or into the reference interval
   — *N.B.* this test is not completely reliable, but may have some therapeutic as well as diagnostic value

Familial hypocalciuric hypercalcaemia

This is a fairly recently described condition:

1. Family history of hypercalcaemia may be present
2. Usually only mild hypercalcaemia

3. Urinary calcium excretion is very low
4. PTH levels may be inappropriately high for the serum calcium
5. Usually a benign condition
   — renal stone disease is not a feature
   — pancreatitis may be a complication

HYPOCALCAEMIA

N.B. Further discussion assumes that the hypocalcaemia is still present after correction for serum albumin level (see 'Physiology' above)

Tetany is the main feature of hypocalcaemia. However, tetany may also occur with alkalosis and a normal total serum calcium—this is because alkalosis reduces the proportion of the total which is ionised, and it is a low ionised calcium which causes tetany.

Causes of hypocalcaemia

1. Osteomalacia/rickets (see 'Metabolic Bone Disease' below)
2. Hypoparathyroidism (see below)
3. Renal failure (see page 22)
4. Acute pancreatitis (see page 55)
5. Hypomagnesaemia (see page 14)

Hypoparathyroidism

Causes:
1. Surgical damage (after thyroid or parathyroid surgery)
2. Idiopathic

Biochemical findings:
1. Hypocalcaemia
2. Hyperphosphataemia
3. Unmeasurable plasma PTH

Pseudohypoparathyroidism:
1. A rare inherited disorder (X-linked inheritance)
2. Clinical presentation similar to idiopathic hypoparathyroidism but the patients have other stigmata such as short metacarpals and mental deficiency
3. Biochemical findings are also similar except that the plasma PTH is high
4. There is renal end-organ resistance to PTH. PTH normally acts by stimulating adenyl cyclase in the end-organ, so that cyclic AMP production is increased. In this condition urinary cyclic AMP fails to rise in response to injected PTH thus differentiating it from hypoparathyroidism.

METABOLIC BONE DISEASE

Includes:
1. Osteoporosis
2. Osteomalacia/rickets
3. Paget's disease (osteitis deformans)
4. Hyperparathyroidism (see page 91)
5. Renal osteodystrophy (see page 23)

Definition of osteoporosis versus osteomalacia

Osteoporosis: reduction in the amount of bony tissue relative to the volume of anatomical bone

Osteomalacia: reduction in the mineralisation of the bony tissue
N.B. the volume of bony tissue relative to the volume of anatomical bone is generally normal or even increased, but its mineral content is reduced

Very occasionally osteoporosis and osteomalacia may occur together.

Osteoporosis
1. Primary osteoporosis
   — post-menopausal/senile osteoporosis
2. Secondary osteoporosis *sometimes* occurs with:
   (i) Cushing's syndrome and steroid therapy
   (ii) Immobilisation
   (iii) Hyperthyroidism
   (iv) Hyperparathyroidism
   (v) Alcoholism
   (vi) Hypogonadism
   (vii) Multiple myeloma } may cause generalised osteoporosis as well as focal lesions
   (viii) Bone metastases } may cause generalised osteoporosis as well as focal lesions
   (ix) Malabsorption states (osteomalacia more common)

Aetiology
Not fully understood but in most cases it seems to be related to increased bone resorption.

Biochemical findings
Normal serum levels of calcium, phosphate and alkaline phosphatase *in primary osteoporosis*.

Osteomalacia/Rickets
See definition of osteomalacia above.

Rickets is the same condition occurring before the epiphyses have closed and therefore leading to a somewhat different clinical presentation.

Aetiology

Deficient vitamin D activity:

1. Vitamin D deficiency
   - (i) inadequate exposure to sunlight
   - (ii) dietary deficiency
   - (iii) malabsorption
2. Vitamin D resistance
   - (i) Impaired production of active vitamin D metabolites:
     - a. chronic renal failure (see 'renal osteodystrophy')
       — impaired renal production of 1,25-DHCC
     - b. anticonvulsant osteomalacia
       — low circulating levels of 25-HCC
       — induction of hepatic microsomal enzymes is thought to lead to enhanced metabolism of vitamin D to inactive metabolites
     - c. chronic liver disease
       — impaired hepatic production of 25-HCC
     - d. vitamin D dependent rickets
       — rare recessively inherited condition
       — may be caused by failure of 1-hydroxylation of 25-HCC by the kidney
   - (ii) Resistance to vitamin D metabolites:
     - a. familial hypophosphataemic rickets (also called 'phosphaturic rickets' or 'vitamin D resistant rickets')
       — sex-linked dominant inheritance
       — failure of renal tubular reabsorption of phosphate
       — marked hypophosphataemia
     - b. chronic acid-base disturbances
       — renal tubular acidosis (see page 29) is the most important example
     - c. Fanconi syndrome (see page 28)

Biochemical findings

1. Serum alkaline phosphatase
   — almost always elevated
2. Serum calcium
   — usually low or low normal
3. Serum phosphate
   — usually low or low normal. Hypophosphataemia tends to be inversely proportional to the serum calcium (probably because when secondary hyperparathyroidism maintains the serum calcium, it causes hypophosphataemia by its phosphaturic action)
   — marked hypophosphataemia occurs in familial hypophosphataemic rickets and Fanconi syndrome

Paget's disease (osteitis deformans)
Aetiology is not known.
Affects part or the whole of one or many bones—never the entire skeleton.
Predominantly a disease of the older age groups.

Biochemical features
1. Raised serum alkaline phosphatase
   — may be very high if many bones are affected by active disease
   — rapidly increasing level suggests the possible development of osteogenic sarcoma
2. Normal serum calcium and phosphate
   — unless the patient is immobilised, when hypercalcaemia may occur

Inborn errors of metabolism

INTRODUCTION

Most inherited biochemical abnormalities can be explained by defective synthesis of a single peptide. The defective peptide may prevent the normal function of the molecule as an enzyme or transport protein. In the classical inborn errors, the affected enzyme is situated on a metabolic pathway. In the lysosomal storage disorders, the defective enzyme is one of a range of enzymes concerned with breakdown of large complex molecules.

The consequences of the defective peptide may be life-threatening or may lead to mental retardation and *early diagnosis is therefore important*. Ante-natal diagnosis in known risk groups is now possible for many enzyme defects (see below).

All inborn errors are rare:

e.g.	
congenital adrenal hyperplasia cystinuria phenylketonuria Hartnup disease	1:10 000–1:25 000
galactosaemia	1:50 000–1:100 000
maple-syrup urine disease tyrosinaemia	1:200 000–1:400 000

Potential consequences of genetic defects

1. The *substrate* of the defective reaction will *accumulate*, e.g.
   (i) phenylalanine in phenylketonuria (see page 104)
   (ii) galactose in galactosaemia (see page 106)
   (iii) 17-hydroxyprogesterone in congenital adrenal hyperplasia (see page 83)
2. There may be *excessive production of other substances* derived from these substrates through alternative pathways, e.g.
   (i) phenylpyruvic acid in phenylketonuria
   (ii) galactitol (→ cataract) in galactosaemia
   (iii) androgens in congenital adrenal hyperplasia
   (iv) lactate in glycogen storage disease type 1 (see page 107)

3. The *product* of the defective reaction may be *reduced* in quantity or even absent, e.g.
   (i) cortisol deficiency in congenital adrenal hyperplasia
   (ii) hypoglycaemia in glycogen storage disease type I
4. Metabolites derived exclusively from the product will also be deficient e.g. melanin deficiency in phenylketonuria
5. When the abnormally accumulating substances are *soluble* they will appear in body fluids and be excreted in increased amounts in the urine (assuming they exceed the renal threshold) e.g. phenylalanine in phenylketonuria
6. When the abnormally accumulating substances are *insoluble* they may be stored, resulting in conditions such as glycogen storage diseases

Normal:

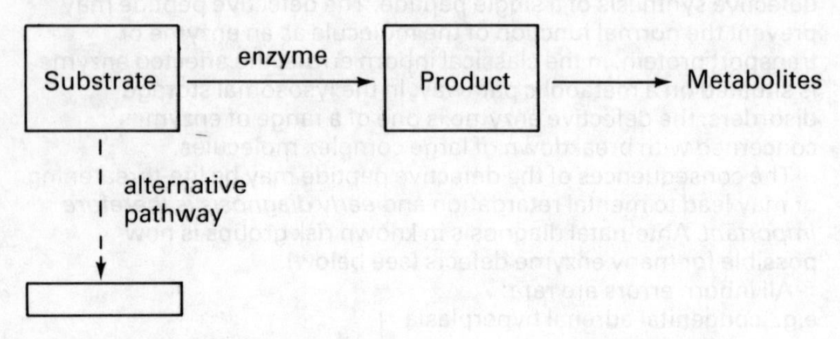

Genetic defect:

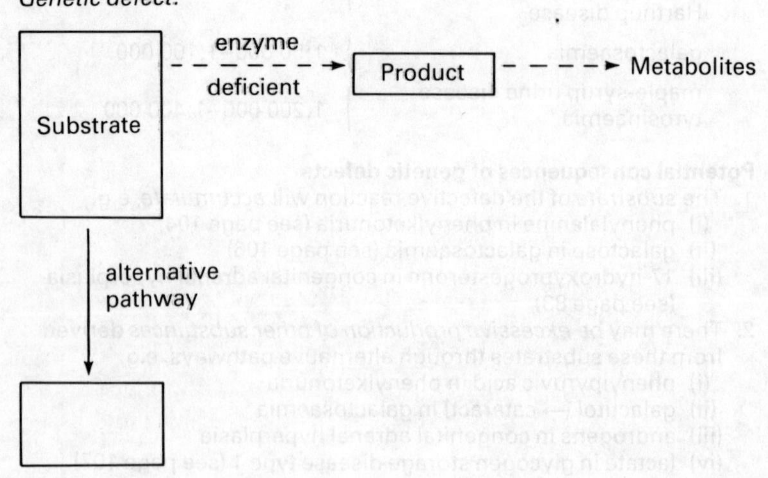

Findings which should suggest an inborn error of metabolism

Disorders which have a very characteristic clinical presentation in later life may present in a non-specific manner in the newborn. Any abnormal clinical or biochemical picture in infancy or childhood, especially in the presence of infection, should alert the physician to the possibility of an inherited metabolic defect.

1. failure to thrive
2. poor feeding
3. persistent vomiting
4. abnormal smell or staining of napkins
5. unexplained hypoglycaemia
6. metabolic acidosis
7. ketosis
8. neurological problems
   - (i) lethargy
   - (ii) loss of suck and swallow reflexes
   - (iii) convulsions
9. liver abnormalities
   - (i) prolonged jaundice
   - (ii) hepatomegaly
   - (iii) acute liver failure
10. renal calculi
11. resistant rickets
12. abnormal hair
13. cataract
14. hyponatraemia/hyperkalaemia
15. neutropenia
16. thrombocytopenia
17. onset of illness related to:
    - (i) change of feeding practice
    - (ii) infection
    - (iii) vaccination
18. similar incidents or unexplained death in previous infant in the family
19. consanguinity in the family

Clinical significance of inborn errors of metabolism

1. In some conditions early diagnosis is imperative as treatment may prevent death or disablement, e.g.
   - (i) phenylketonuria } control by diet
   - (ii) galactosaemia } control by diet
   - (iii) congenital hypothyroidism—replace thyroxine
   - (iv) glycogen storage disease type I—treat by frequent feeding
2. In some instances studies should be carried out in order to identify relatives at risk and preventive measures instituted where necessary, e.g.

(i) hepatic porphyrias (see page 110)
— acute attacks precipitated by a number of drugs
(ii) cholinesterase abnormalities (see page 165)
— suxamethonium sensitivity
(iii) cystinuria
— formation of calculi
(iv) genetically inherited haemochromatosis (see page 120)
— accumulation of iron
(v) Wilson's disease (see page 126)
— accumulation of copper

3. Some inborn errors are of little consequence in themselves but may cause concern or confusion when discovered accidentally, e.g.
   (i) Gilbert's syndrome (see page 47)
   (ii) renal glycosuria

Genetic considerations

Inheritance

— *In general terms*:

1. Most enzymes are the product of autosomal genes. Normally an individual will inherit one gene from the mother and one from the father, so 50% of that individual's enzyme will be the product of the maternal gene and 50% the product of the paternal gene.
2. If a defective gene is such that either there is no gene product or the protein produced has no enzyme activity, then an individual heterozygous for that gene will have 50% of a normal individual's enzyme activity. An individual homozygous for that gene will have no enzyme activity.
3. In other situations, an abnormal gene may code for a molecule with some enzyme activity, but less than normal. An individual heterozygous for that gene will have enzyme activity somewhere between 50–100% of normal.
4. The degree of deficiency of enzyme activity necessary to produce clinical disease varies from enzyme to enzyme.
   (i) If an inherited metabolic disorder is such that heterozygotes for the abnormal gene have enzyme activities below the level which produces clinical disease, the disorder is said to have *dominant* inheritance e.g. acute intermittent porphyria or familial hypercholesterolaemia. The rare homozygotes are more severely affected than heterozygotes.
   (ii) If the disorder is such that only homozygotes for the abnormal gene have clinical disease, the inheritance is said to be *recessive*. Most of the classical inborn errors of metabolism fall into this group.
5. In the case of genes carried on the X chromosome, inheritance is said to be X-linked.
   (i) In males, the only X chromosome is that derived from the mother.

(ii) In females, one X chromosome is derived from the mother and one from the father, but in any cell, only one X chromosome is active, the other being converted to an inactive Barr body. According to the so-called *Lyon hypothesis,* this inactivation occurs randomly, so that on average 50% of maternal X chromosomes and 50% of paternal X chromosomes will be active.

(iii) From (ii) it follows that females heterozygous for an abnormal gene will always have at least 50% of normal gene product levels. In most X-linked disorders such females are not clinically affected and only males and homozygous females are affected—this is *X-linked recessive* inheritance e.g. haemophilia, glucose-6-phosphate dehydrogenase (G6PD) deficiency.

(iv) In one or two disorders, heterozygous females are affected, though less severely than affected males and homozygous females—this is *X-linked dominant* inheritance e.g. familial hypophosphataemic rickets.

— *With autosomal recessive inheritance*:
1. The patient is homozygous (male or female)
2. Both parents are heterozygous carriers
3. Following the birth of an affected child, the chance of a subsequent sibling being affected is 1 in 4

— *With X-linked recessive inheritance*:
1. The patient is male (except for rare homozygous females)
2. The mother is a heterozygous carrier
3. Half her male offspring will be affected
4. The female offspring will virtually always be healthy *but*
5. Half the female offspring will be carriers

N.B. With dominant inheritance patients generally appear in successive generations (except when there is incomplete penetrance) whereas with recessive inheritance the parents almost always appear normal.

Screening
1. Screening for affected individuals
   (i) Whole population
      a. phenylketonuria in the UK
      b. galactosaemia } possible but not applied at
      c. G6PD deficiency } present in UK
   (ii) Selected population
      a. sick children in hospital
      b. occasionally adults known to be at high risk e.g. Jews for Tay-Sachs disease
      c. siblings of known cases (pre or post-natal)

2. Screening for carriers
   N.B. there is often no *sure* way of demonstrating heterozygotes
   (i) Determine enzyme activity
   — in a few instances heterozygote values can be distinguished from those of a normal population e.g. leucocyte hexosaminidase A in Tay-Sachs, erythrocyte galactose-1-phosphate uridyl transferase activity in galactosaemia.
   (ii) Stress the defective enzyme system with a load of its normal substrate to reveal:
   — increased and prolonged blood concentration
   — excretion of unusual metabolites
   e.g. phenylalanine and phenolic acids in phenylketonuria.
   (iii) Electrophoretic methods may be used for haemoglobinopathies and (e.g. protein abnormalities alpha-1-antitrypsin deficiency)

Ante-natal diagnosis

Ante-natal diagnosis of specific enzyme defects in situations where
1. the diagnosis has already been made in a previously affected offspring
2. the parents have been shown to be heterozygous or the mother is a known carrier of an X-linked disorder

— *Investigations*:
1. Examination of uncultured cells for sex (X-linked disorders only)
   female—health assured
   male—risk is 1 in 2
2. Examination of cultured cells for the enzyme defect

Counselling

It is most important that information given to parents and other relatives of an affected child is based on results from the most reliable laboratory methods available. As such information will be passed to successive generations in the extended family, all diagnosis of inborn errors should be reviewed from time to time in the light of technological advances.

Vitamin responsive variants

Abnormal enzyme structure in the region of the binding site can affect activity through a change in affinity of the site for a co-enzyme. In many cases a vitamin is the precursor of the co-enzyme and the diminished enzyme activity may then be enhanced by giving larger amounts of the vitamin e.g. there is a variant of maple syrup urine disease which is responsive to thiamine.

Laboratory investigation

The findings which should suggest the need for an 'inborn error screen' are given on page 99.

Routine tests (e.g. glucose, electrolytes, acid/base, bilirubin) may provide useful information.

Samples: often require special collection/storage requirements — consult the laboratory *before* collection

Screening tests

— *Urine*:

1. Reducing substances ('Clinitest') e.g. for galactose, fructose. If positive → sugar chromatography for identification.
2. Cystine screen—for cystine and homocystine
3. Aminoacid chromatography (*N.B.* blood should be examined as well)

— *Stool*:

Sugar chromatography is important as an exclusion test in infants with 'failure to thrive' who might have a primary disaccharidase deficiency (see page 57)

— *Blood*:

1. Screening test for erythrocyte galactose-1-phosphatase uridyl transferase (see galactosaemia below)—important in the differential diagnosis of prolonged jaundice.
2. Aminoacid chromatography
3. Ammonia
   — when a urea cycle disorder or organic aciduria is suspected
4. Lactate
   — when there is a metabolic acidosis
   — important in:
   (i) some disorders of carbohydrate metabolism e.g. von Gierke's disease
   (ii) organic acidurias

Definitive tests

Often direct enzyme assay on:

1. leucocytes
2. erythrocytes
3. biopsy tissue e.g. intestinal mucosa, cultured skin fibroblasts, liver, muscle, thyroid.

Hazards

1. Immature infants—may give transient test abnormalities, e.g.
   (i) galactose-1-phosphate uridyl transferase deficiency may be found in premature infants as well as in galactosaemia (*false positive*)

(ii) tyrosine or phenylalanine may accumulate in the blood due to immaturity of liver enzymes (*false positive*)

2. Diet
   (i) if tests are performed before a full diet is instituted, metabolites may not have had time to accumulate e.g. in phenylketonuria before the 6th day of life (*false negative*)
   (ii) infants on clear fluids may not demonstrate the abnormality under investigation (*false negative*)

AMINOACID DISORDERS

The following conditions are illustrative of the types of aminoacid disorder which may occur.

Phenylketonuria

Clinical features:
1. irritability and vomiting (early weeks)
2. mental retardation (4–6 months)
3. fair complexion (deficient melanin—see below)

Aetiology:
Enzyme deficiency (phenylalanine hydroxylase)

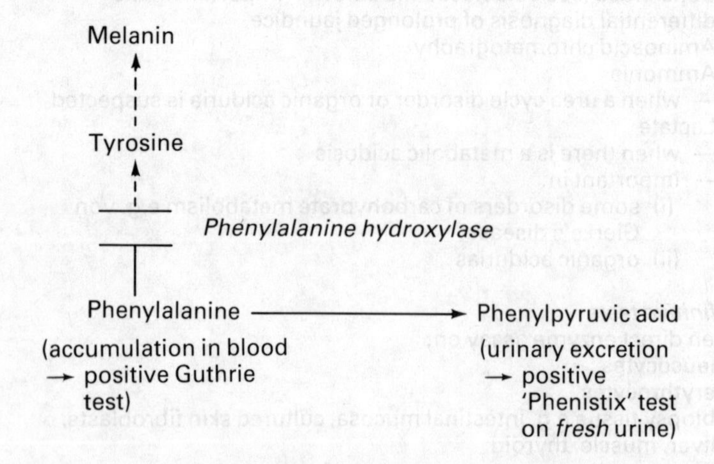

N.B. it is the accumulation of phenylalanine which leads to cerebral damage

Treatment:
1. Diet low in phenylalanine
2. Monitor blood phenylalanine levels to assess efficacy of diet
3. Diet can be discontinued in adult life when brain development is complete *but*:

   N.B. high phenylalanine levels in a phenylketonuric pregnant mother will result in a microcephalic infant (even if the latter does not have phenylketonuria)—mother must resume diet for the duration of pregnancy.

Cystinuria

Clinical features:
1. often asymptomatic
2. renal calculi (consisting of cystine) in homozygotes
3. increased cystine excretion in heterozygotes

Aetiology
Transport defect
Failure of the mechanism for the renal tubular reabsorption of the group of aminoacids which comprises cystine and the basic aminoacids (ornithine, arginine and lysine) (Mnemonic: 'COAL')
Urine shows a positive cystine screen and typical aminoacid pattern.

Treatment:
— avoid dehydration

Hartnup disease

Clinical features:
1. pellagra-like rash
2. ataxia
3. confusion

} may be intermittent

Aetiology:
Transport defect
Failure of the mechanism for the intestinal absorption and renal tubular reabsorption of neutral aminoacids (in particular tryptophan).
The symptoms could be related to wastage of tryptophan, the precursor of nicotinic acid i.e. vitamin B_2 (deficiency of which causes pellagra).
There is a typical urine aminoacid pattern.

Treatment:
— symptoms alleviated by nicotinamide

Cystinosis

Clinical features:
1. Fanconi syndrome (see page 28)
2. Progressive renal failure
3. Photophobia

Aetiology
Presumed enzyme deficiency
Accumulation of cystine in nephrons (→ Fanconi syndrome and renal failure) and the retina (→ photophobia).

Aminoaciduria
Many inborn errors of aminoacid metabolism lead to aminoaciduria

Classification:
1. Generalised aminoaciduria
   (i) overflow
   — high blood levels of many aminoacids as in liver failure
   — not caused by inborn errors of aminoacid metabolism.
   (ii) renal tubular disorder (see page 28)
   — Fanconi syndrome as in cystinosis, galactosaemia, heavy metal poisoning etc.
   — vitamin D deficiency
2. Specific aminoaciduria
   (i) overflow
   — high blood level related to an inborn error e.g. phenylalanine as in phenylketonuria
   (ii) renal tubular transport defect
   — e.g. cystinuria and Hartnup disease (see above)

CARBOHYDRATE DISORDERS

Galactosaemia

Clinical features:
1. hypoglycaemia
2. prolonged jaundice in the neonate (> 10 days)
3. slowness to feed and vomiting
4. transient hepatomegaly
5. mental retardation } 2–3 years of age
6. cataracts }
7. Fanconi syndrome (see page 28)

Aetiology:
enzyme deficiency (galactose-1-phosphate uridyl transferase)

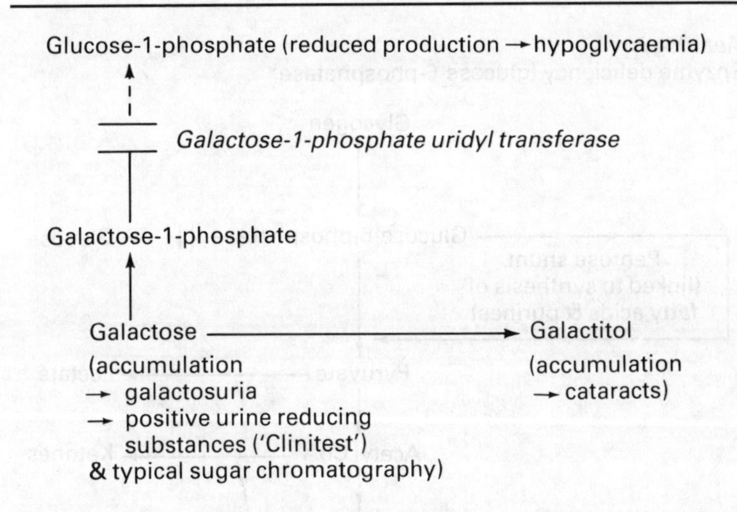

N.B. Clinical features will not appear until milk has been included in the diet.

and:

Vomiting → withdrawal of milk feeds → disappearance of galactose from urine → negative urine tests

Definitive diagnosis is by measurement of erythrocyte galactose-1-phosphate uridyl transferase activity.

Treatment:
— galactose-free diet (i.e. eliminate milk and milk products from the diet)
— if galactosaemia is suspected (even in the absence of galactosuria) a galactose-free diet should be instituted until an erythrocyte galactose-1-phosphate uridyl transferase assay result is available.

Glycogen storage disease type I (von Gierke's disease)

Clinical features:
1. hypoglycaemia
2. hepatomegaly (caused by glycogen storage)
3. ketosis
4. lactic acidosis
5. hypertriglyceridaemia
6. hyperuricaemia

Aetiology:
Enzyme deficiency (glucose-6-phosphatase)

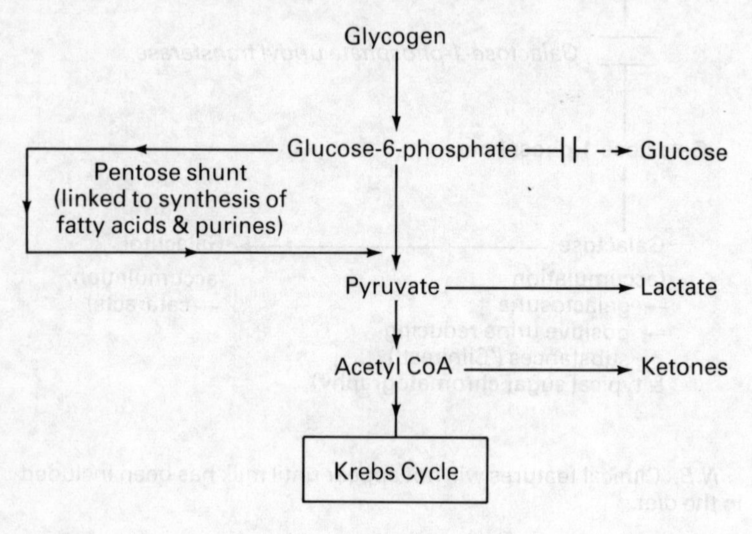

Treatment:
— frequent feeding to maintain blood glucose

Porphyria

PORPHYRIN METABOLISM

Porphyrinogens (the hexahydro derivatives of porphyrins) and protoporphyrin IX are involved in the biosynthetic pathway of haem.

Haem biosynthesis

Porphobilinogen (PBG) is formed from two molecules of 5-aminolaevulinate (ALA) which in turn is derived from glycine and succinyl co-enzyme A in a reaction catalysed by ALA synthetase. There is normally a negative feedback from the final product (haem) to control ALA synthetase. Four molecules of PBG are then polymerized to form one of the two isomers of uroporphyrinogen — uroporphyrinogen III production for the main haem pathway requiring the action of two enzymes, whereas uroporphyrinogen I production requires only one.

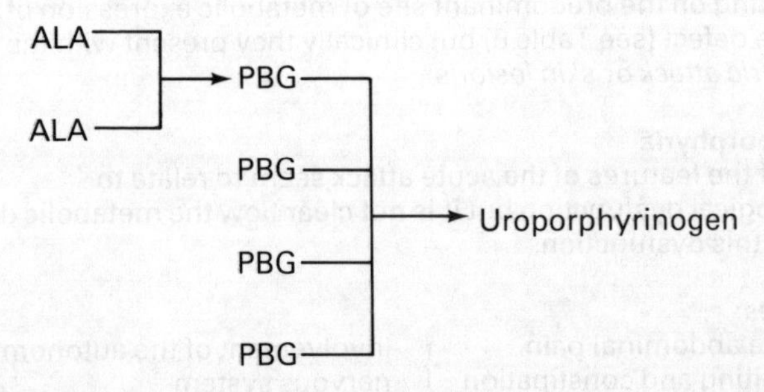

The full haem biosynthetic pathway with the enzymes involved is shown in Fig. 15.

Excretion of haem precursors

— At all stages in the biosynthetic pathway the porphyrinogens are readily converted to the corresponding porphyrins which, with the exception of protoporphyrin, cannot be further metabolised.
— In moving along the pathway the porphyrins become decreasingly polar i.e. less water soluble.

— Haem precursors normally appear in small amounts in urine and in faeces (the latter through biliary excretion), their major route of excretion depending on their polarity (see Fig. 15)
— Urine screening tests are not sensitive enough to detect such small amounts, but faecal porphyrin screens may show faint fluorescence. (Screening tests for porphyrins depend on the property of fluorescence which porphyrins show under certain conditions).
— The excretion of coproporphyrin transfers from faeces to urine if hepatic excretion is impaired.

THE PORPHYRIAS

The porphyrias are a group of disorders, each of which results from an inborn error of one of the enzymes involved in haem synthesis (but see 'porphyria cutanea tarda symptomatica' below). The enzymes thought to be defective are shown in Table 5.

The characteristic clinical and biochemical findings are related to reduced activity of the particular defective enzyme. The activity of ALA synthetase increases because of decreased negative feedback from haem and this usually results in normal haem production but at the expense of a build-up of porphyrin and/or porphyrin precursors prior to the block. The pattern of this accumulation can be used to distinguish one type of porphyria from another. Measurement of specific enzyme activity is not readily available but may be necessary to detect latent cases in family studies.

The porphyrias are often divided into *hepatic* and *erythropoietic* depending on the predominant site of metabolic expression of the enzyme defect (see Table 6) but clinically they present with the *acute porphyric attack* or *skin lesions*:

Acute porphyria

Most of the features of the acute attack seem to relate to neurological dysfunction but it is not clear how the metabolic defect causes this dysfunction.

Features:

1. acute abdominal pain } involvement of the autonomic
2. vomiting and constipation } nervous system
3. psychiatric disturbance
4. other neurological dysfunction (peripheral neuritis through to death from respiratory paralysis)
5. hyponatraemia (related to vomiting, poor renal function or inappropriate ADH secretion)
6. urine which is reddish-brown (or may only darken on standing)

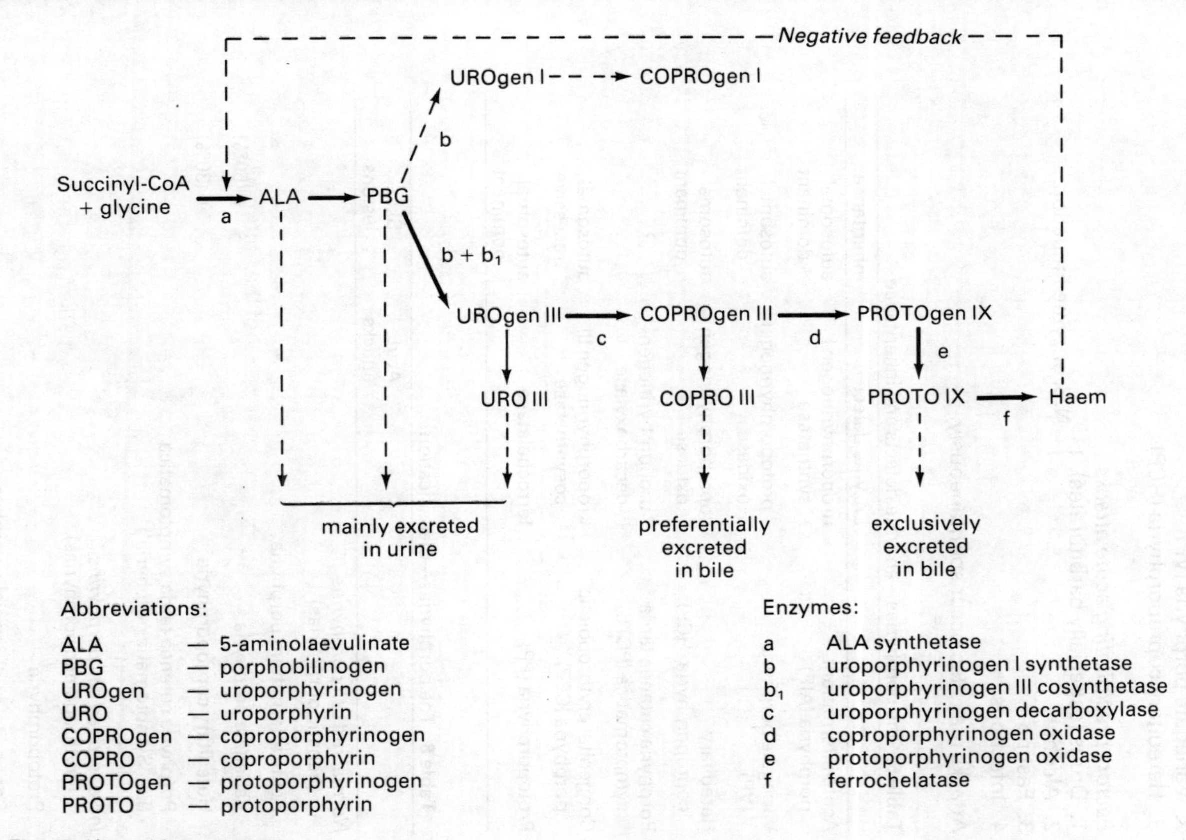

Fig. 15 Haem biosynthesis

Acute attacks may occur in:
1. Acute intermittent porphyria (AIP)
2. Variegate porphyria (VP)
3. Hereditary coproporphyria (HCP)

Factors precipitating acute attacks
1. Drugs (especially barbiturates) } N.B. enzyme induction
2. Alcohol }
3. Fasting
4. Infections

Attacks may also occur spontaneously.

Table 5. The porphyrias—enzyme defects and inheritance

	Enzyme defect	Inheritance
Acute intermittent porphyria (AIP)	uroporphyrinogen I synthetase	autosomal dominant
Variegate porphyria (VP)	? protoporphyrinogen oxidase	autosomal dominant
Hereditary coproporphyria (HCP)	coproporphyrinogen oxidase	autosomal dominant
Porphyria cutanea tarda symptomatica (PCTS)	? uroporphyrinogen decarboxylase	?
Congenital erythropoietic porphyria (CEP)	uroporphyrinogen III cosynthetase	autosomal recessive
Protoporphyria (PP)	ferrochelatase	autosomal dominant

Table 6. The porphyrias—Classification

	Acute attacks	Skin lesions
Normal red cell porphyrins (hepatic porphyrias)		
Acute intermittent porphyria	+	–
Variegate porphyria	+	+ (50%)
Hereditary coproporphyria	+	+ (30%)
Porphyria cutanea tarda symptomatica (i.e. 'Symptomatic porphyria')	–	+
Increased red cell porphyrins (erythropoietic porphyrias)		
Protoporphyria	–	+
Congenital erythropoietic porphyria	–	+

Metabolic basis of the acute attack
Acute attacks seem to relate to induction of ALA synthetase.

Many of the drugs that provoke acute attacks are metabolised in the liver by haem-containing microsomal enzyme systems. These drugs normally induce a short-lived increase in ALA synthetase appropriate to the production of the extra haemoprotein (particularly cytochrome P_{450}) required for their metabolism. However, in acute porphyria, the increase in ALA synthetase is sustained because the defective enzyme impairs the ability to increase haem production sufficienty. This leads to massive overproduction of the intermediates preceding the block in the pathway.

N.B. acute attacks only occur in those porphyrias which are associated with accumulation of porphyrin *precursors*. They do *not* occur if the excretion of PBG is always within normal limits.

(It has been suggested that the factor which determines whether PBG accumulates in a particular form of porphyria is the uroporphyrinogen I synthetase activity. Its activity appears to be increased in the forms of porphyria which are not associated with acute attacks.)

Screening tests in acute porphyria (see also Fig. 16)
1. Urine
   (i) Porphobilinogen (PBG) screen
   — *positive* in all three acute porphyrias *during an acute attack*
   — only becomes positive at levels above 3 × the upper limit of normal.
   — in remission, may be negative in AIP and will usually be negative in VP and HCP.
   (ii) Porphyrin screen
   — usually positive in all three acute porphyrias during an acute attack (in AIP because of non-enzymic polymerisation of PBG to form uroporphyrin)

N.B. abdominal pain
positive urine porphyrin
negative urine PBG } usually secondary coproporphyrinuria *not* porphyria

Causes of secondary coproporphyrinuria:
a. alcoholism
b. lead poisoning
c. liver disease
d. leukaemia
e. Hodgkin's
f. aplastic anaemia
g. pernicious anaemia
h. haemolytic anaemia

During an attack:

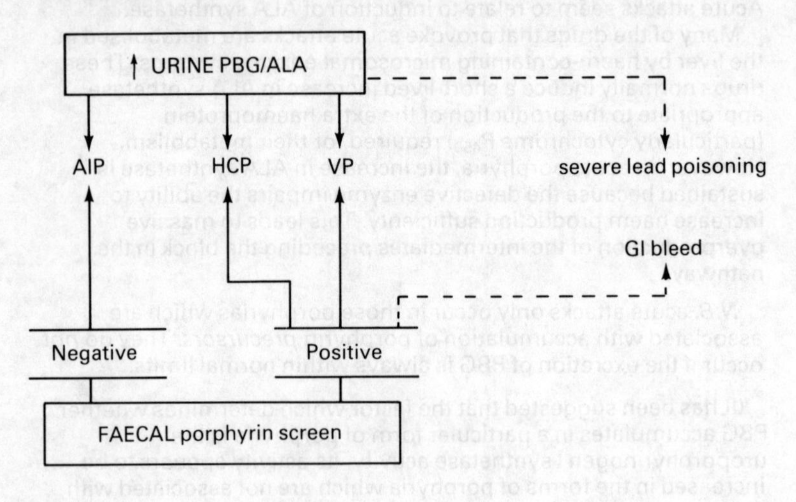

In remission:

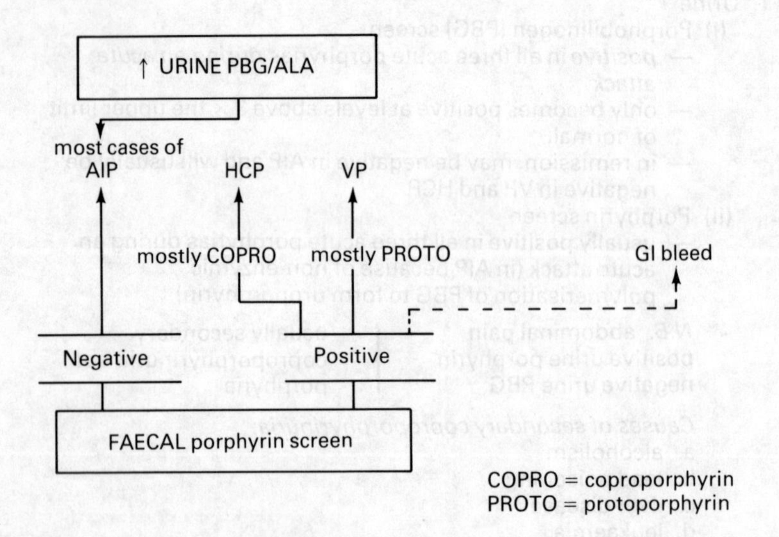

COPRO = coproporphyrin
PROTO = protoporphyrin

Fig. 16. Tests in acute porphyria

2. Faeces
   (i) Porphyrin screen for:
      a. *detection* of VP and HCP when PBG excretion is normal
      b. *differentiation* of:
         AIP—normal excretion
         VP } increased excretion*
         HCP }
         * discriminate further with quantitative porphyrins
   (ii) Occult blood will help to elucidate possible false positive faecal porphyrin screen caused by gastrointestinal bleeding (haem converted to porphyrin by intestinal bacteria)

Cutaneous porphyrias

Features:
1. photosensitivity
2. increased fragility of the skin
3. subepidermal bullae
4. oedema (sun sensitive)
5. pigmentation
6. hirsutism

Skin lesions may occur in:
1. porphyria cutanea tarda symptomatica (PCTS)—also called 'symptomatic porphyria'
2. variegate porphyria (VP) } acute attacks also occur (see
3. hereditary coproporphyria (HCP) } above)
4. protoporphyria (PP)
5. congenital erythropoietic porphyria (CEP)—very rare

Metabolic basis of skin lesions
Sensitivity to sunlight is the result of the photodynamic action of the porphyrins that accumulate in the skin when the plasma porphyrin concentration is increased.

Skin lesions do not occur in AIP because in that condition only porphyrin precursors accumulate, not porphyrins.

Porphyria cutanea tarda symptomatica (PCTS)
This deserves special attention because it is the commonest form of cutaneous porphyria in the United Kingdom and it is atypical in terms of inheritance.

It has been regarded as acquired, but it seems likely that it only occurs in genetically predisposed individuals. It is often associated with liver disease (particularly when due to alcohol).

Hepatic iron overload is very common in PCTS and depletion of storage iron by repeated venesection produces a clinical remission in most patients.

The metabolic defect in this condition appears to involve the conversion of uroporphyrinogen III to coproporphyrinogen III, a process which involves several intermediate steps. Uroporphyrinogen III accumulates, but also other intermediates which lead to the excretion of unusual porphyrins (e.g. isocoproporphyrin in the faeces—this appears to be specific for PCTS)

Screening tests in cutaneous porphyria

1. Blood
   — increased erythrocyte porphyrin:
   (i) PP
   (ii) CEP
   — deficiency of uroporphyrin III cosynthetase → increased production and build-up of porphyrins of isomer series I
2. Urine
   — increased urine porphyrin:
   (i) PCTS (uroporphyrin)
   (ii) HCP (coproporphyrin)
   (iii) sometimes in VP (coproporphyrin)
   (iv) CEP (uroporphyrin I and some coproporphyrin I)

   N.B. Secondary coproporphyrinuria is also a common cause of increased urine porphyrin (see page 113)
3. Faeces
   — increased faecal porphyrin:
   (i) VP (proto-and coproporphyrin)
   (ii) HCP (coproporphyrin)
   (iii) PP (proto- and sometimes coproporphyrin)
   (iv) CEP (coproporphyrin I)
   (v) PCTS (unusual porphyrins—see 'PCTS' above)

Iron and ferritin

PHYSIOLOGY

The average male adult has a total body iron of approximately 4 g:

3 g	haemoglobin	(oxygen transport)
	myoglobin	(oxygen storage)
	enzymes	
1 g	ferritin	iron storage (see below)
	haemosiderin	

Since free iron is toxic it circulates bound to the carrier protein transferrin and is stored as ferritin (bound to the protein apoferritin).

Regulation

Iron balance in the body is controlled by absorption (the principal site of absorption being the upper small intestine). There is very little iron excretion.

Apoferritin

A proportion of the dietary iron binds to apoferritin in the intestinal columnar epithelium and is later shed with it. The remainder is actively transported through the intestinal mucosa and is distributed by transferrin to the marrow for developing erythrocytes and to various sites for storage.

When iron stores are deficient there is only a little apoferritin in the epithelium and most of the iron is absorbed. In iron overload, the amount of apoferritin is greatly increased so that less iron is absorbed into the circulation.

Transferrin

Transferrin is normally one third saturated with iron. When iron reserves diminish and the demand for iron increases, the serum transferrin level rises and the life of the protein is extended. The abundance of unsaturated transferrin restricts the storage of iron and delivers it preferentially to the marrow for erythropoiesis. In iron overload, the transferrin is reduced and iron is diverted to store.

Factors affecting the serum iron concentrations in health

1. Age
   — serum iron levels are lower during the first two years of life
2. Sex
   — levels are lower in women than men between puberty and the menopause (hormonal influence rather than menstrual loss)
3. Circadian rhythm (may be up to 50%)
   — highest 0700–1000 hours
   — lowest 1700–2000 hours
4. Day to day variation
   — also quite marked
   — levels may be very low just before or during menstruation

Iron stores

Iron is stored in:

1. Reticuloendothelial (RE) cells (bone marrow, spleen, liver)
2. Hepatic parenchymal cells
3. Skeletal muscle (low concentration but large mass; iron not readily available)

The storage compounds are ferritin and haemosiderin (which is probably a ferritin aggregate). The iron in ferritin is more readily available.

RE cell iron comes largely from phagocytosis of ageing or defective erythrocytes. The RE cell stores are the most active in internal iron exchange and require the presence of vitamin C to release iron to transferrin (so in vitamin C deficiency serum iron may be low even though the RE cell stores of iron are high).

A small amount of ferritin normally circulates in serum and in most circumstances, this level reflects the body iron stores.

IRON DEFICIENCY

Iron depletion

Decreased iron stores without clinical anaemia may occur in:

1. Growth spurts in children } increased requirement
2. Pregnancy }
3. Persistent slight gastrointestinal bleeding—patient unaware e.g. with salicylate use

Iron deficiency anaemia

Aetiology:

1. Deficient intake (difficult to achieve)
   — in general, vegetable iron is less readily absorbed than that from animal sources.

2. Defective absorption, e.g.
   (i) coeliac disease
   (ii) partial gastrectomy
   (iii) jejunal resection
3. Chronic blood loss (probably > 15 ml/day)
   (i) gastrointestinal bleeding e.g. haemorrhoids, peptic ulcer, malignancy
   (ii) excessive menstrual loss
   (iii) hookworm (in endemic areas)

Differential diagnosis:
1. Anaemia of chronic disease, e.g.
   (i) chronic infection
   (ii) malignancy
   (iii) renal failure
2. Anaemia associated with thalassaemia.(rare)
3. Pyridoxine-responsive anaemia (rare)

Laboratory investigation:
1. Haematological
   — usually diagnostic
2. Biochemical
   — not indicated in simple iron deficiency but only in difficult cases or when an anaemia could be of dual aetiology
   (i) Serum iron
      — not very useful because of marked physiological variation
      — low in iron deficiency and anaemia of chronic disease
   (ii) Serum transferrin or total iron-binding capacity
      a. *increase* in iron deficiency
      b. *decrease* in:
         — anaemia of chronic disease
         — iron overload
         — nephrotic syndrome (renal loss)
   (iii) Serum ferritin
      — generally reflects total body iron stores except in:
      a. hepatocellular damage (releasing ferritin)
      b. leukaemia
      c. Hodgkin's disease

IRON OVERLOAD

1. Reticuloendothelial (bone marrow, spleen, liver)
   May be due to:
   (i) repeated blood transfusions
   (ii) overenthusiastic parenteral iron therapy (especially if there is a misdiagnosis of 'iron deficiency')
   (iii) dietary (rare)

2. Parenchymal (liver, pancreas, heart)
   May be due to:
   (i) idiopathic haemochromatosis
   (ii) ineffective erythropoiesis
   (iii) inappropriate oral iron
   (iv) dietary } rare
   (v) cirrhosis } rare
   (vi) portocaval shunt } rare

N.B. When iron overload becomes sufficiently extensive (whatever the cause) there will be parenchymal cell involvement.

Idiopathic haemochromatosis

Aetiology:
Hereditary increase in intestinal absorption of iron, usually presenting in middle life (may be accelerated by alcoholism)

Presents with:
1. skin pigmentation (melanin)
2. liver disease (cirrhosis)
3. diabetes
4. cardiac failure (especially the younger patients)
5. hypogonadism

(2–5: caused by deposition of iron in liver, pancreas, heart and gonads respectively)

Differential diagnosis:
Easily distinguished from most forms of secondary haemochromatosis but some cases of cirrhosis may cause confusion.

Laboratory investigations:
1. First line tests
   (i) serum iron—high
   (ii) transferrin } normal or low
   TIBC } normal or low
   (i and ii: high saturation)
2. Tests of body iron stores
   (i) serum ferritin—high
   — depends on:
   a. degree of liver cell damage
   b. magnitude of hepatic iron stores
   — may be normal in latent disease
   (ii) chelation tests
   — if stores are excessive there is increased urinary excretion of iron after administration of a suitable chelating agent

(iii) liver biopsy
a. may show cirrhosis
b. may show stainable iron present
— in RE cells
and/or
— in parenchymal cells
(iv) bone marrow
— stainable iron usually normal in haemochromatosis (high in RE cell overload)

3. Secondary investigations for:
(i) liver disease
(ii) diabetes
(iii) cardiac disease
4. Family studies
These are worth while because early treatment by venesection is effective. Relatives with HLA haplotypes identical to that of the patient are most likely to develop the disease.

Nutrition

VITAMINS AND TRACE ELEMENTS

Vitamins

Vitamins (other than D and E) function as cofactors, or cofactor precursors, for enzyme systems.

Sub-clinical vitamin deficiency is of particular relevance in the sick (where utilisation may be increased by the metabolic requirements for recovery) and in the elderly.

Assessment of vitamin status

1. Dietary history
2. Clinical examination
3. Laboratory investigation

Best laboratory indices of vitamin status:

Vitamin	
A	plasma retinol and β-carotene
B_1	thiamine pyrophosphate (TPP) effect on erythrocyte transketolase
B_2	flavin adenine dinucleotide (FAD)
B_6	pyridoxine 5-phosphate (PLP) effect on red cell AST
B_{12}	serum B_{12}
Folic acid	red cell folate
C	leucocyte ascorbic acid
D	plasma 25-hydroxycholecalciferol
E	plasma tocopherol
K	prothrombin time

N.B. Clinical responsiveness is often the best test of vitamin status

Deficiency

Deficiency is rare in affluent communities eating a varied diet and single vitamin deficiency is even more rare (except perhaps for B_{12} and folate).

Causes of vitamin deficiency:

1. Restricted dietary intake
   (i) inadequate availability of vitamin-containing foods (especially in the institutionalised and elderly)
   (ii) poor appetite
   (iii) alcoholism
   (iv) obstruction in the upper digestive tract
   (v) fad foods, ethnic minorities
   (vi) inadequately supplemented parenteral nutrition
2. Increased requirement (relative deficiency)
   (i) pregnancy/lactation
   (ii) growth spurts in children
   (iii) infections
   (iv) some drug therapies (e.g. isoniazid and penicillamine affect B_6)
3. Decreased absorption
   (i) gastrointestinal disease
   (ii) fat malabsorption (fat-soluble vitamins affected)
   (iii) blind loop syndrome (bacteria compete with intrinsic factor for B_{12})
4. Impaired synthesis, utilisation or storage
   (i) lack of sunlight → reduced synthesis of vitamin D in the skin
   (ii) renal disease → failure of 1-hydroxylation of 25-OH vitamin D (see page 23)
   (iii) prolonged oral broad spectrum antibiotics → reduced synthesis of vitamin K by colonic bacteria
   (iv) liver disease

Fat-soluble vitamins

Fat-soluble vitamins (A, D, E & K) being insoluble in water are transported in the blood linked to protein and so are not readily lost in the urine. Their absorption depends on normal fat absorption. They are stored in the liver and other tissues, so deficiency takes a considerable time to develop but toxicity from over-ingestion can occur (especially with A and D).

Calciferol (vitamin D)

Vitamin D is discussed in chapter IX.

Vitamin E

Deficiency in infancy → haemolytic anaemia

Vitamin K

The synthesis of prothrombin and clotting factors VII, IX and X is vitamin K dependent. Bleeding may occur when the prothrombin time falls to one third of normal or below.

Water-soluble vitamins

Water-soluble vitamins (B complex and C) are not stored to any great extent (with the exception of B_{12}), any excess intake being excreted in the urine. Short term deficiency can therefore develop, but ill effects from excess ingestion are rare.

Thiamine (Vitamin B_1)

The active form, thiamine pyrophosphate, acts as co-enzyme in a number of metabolic pathways including:

1. entry of carbohydrate into the citric acid cycle
2. metabolism of glucose in red cells and neurological tissues

Low levels are associated with:

1. poverty
2. old age
3. alcoholism (in some cases)
4. excessive dependence on polished rice

Presenting clinical features include:

1. cardiomyopathy
2. polyneuropathy
3. mental changes (e.g. Korsakoff's psychosis in alcoholics)

N.B.

— glucose administration may aggravate the clinical picture by increasing the requirement for thiamine

— without thiamine, pyruvate will accumulate in the blood as it cannot be converted to acetylcoenzyme A.

Ascorbic acid (vitamin C)

Hydroxylation of lysine and proline during collagen synthesis is the only well established role for ascorbic acid, but it may act as a hydrogen carrier throughout the tissues.

Poor collagen formation leads to deficient vascular walls and contributes to many of the clinical features:

1. gingivitis
2. petechial haemorrhages
3. poor healing of wounds/fractures
4. anaemia (multifactorial)
5. haemarthroses } in infants
6. subperiosteal bleeding } in infants

Vitamin C deficiency usually presents nowadays as "bruising easily", most commonly in elderly men living alone in poor circumstances.

Trace elements

Like vitamins, the chief biological function of trace elements is in enzyme systems—either as part of the enzyme molecule (metalloenzymes) or as enzyme co-factors. In plasma these elements are reversibly bound to albumin, and transferrin can bind chromium, copper, manganese and zinc in addition to iron. More than 90% of plasma copper is incorporated in caeruloplasmin and there are alpha-1 and alpha-2 globulins which form complexes with zinc. Plasma levels of trace elements are therefore related in some instances to plasma protein concentrations.

Absorption
Occurs throughout the small intestine

Excretion

urinary	*faecal*	
chromium	copper	through bile
cobalt	manganese	
iodine		
molybdenum	iron	mucosal
selenium	zinc	desquamation

Assessment of trace element status
Blood levels are usually measured; hair is of little help in acute clinical situations but may give a better indication of tissue levels over a longer period. The level of one trace element relative to another may be important and is most easily established in hair samples.

N.B. Levels of trace elements in hair may be affected by dyeing, bleaching or permanent waving.

Deficiency
Although much of the work on trace elements has been in connection with animal husbandry, in man gross deficiencies have been shown to occur of:
— chromium
— copper
— iodine
— iron
— zinc

Causes of trace element deficiency:
1. Restricted dietary intake
   (i) protein restricted diet
   (ii) special diets (casein instead of natural protein) e.g. for phenylketonuria
   (iii) impoverished or institutionalised elderly (cheap vegetable protein)

2. Decreased absorption
   (i) immature absorptive mechanism (pre-term infants)
   (ii) gastrointestinal disease
   (iii) acrodermatitis enteropathica
   (iv) Menke's disease
3. Increased loss
   (i) catabolic states
   (ii) diuretic therapy
   (iii) proteinuria
   (iv) exfoliative dermatitis
   (v) excessive sweating } tropics
   (vi) chronic haemorrhage e.g. intestinal parasites } tropics
4. Total parenteral nutrition especially:
   (i) neonates
   (ii) adults with malabsorption

Copper

Function
Plays an important role in release of iron from body stores and in metabolism and transport of iron within the normoblast. Involved in many enzyme systems.

Clinical syndromes
1. Resistant hypochromic anaemia
   — associated with neutropenia and copper deficiency,
   — usually resulting from artificial diets or total parenteral nutrition (especially in infants)
   — responds to iron only when copper is given as well
2. Inborn errors of metabolism
   (i) Wilson's disease
   — presents with:
      a. symptoms of liver disease
      b. neurological features
   — findings:
      a. excess deposition of copper in
         liver → cirrhosis
         brain → basal ganglia degeneration
         kidneys → Fanconi syndrome
         cornea → Kayser-Fleischer rings
      b. often low levels of serum copper and caeruloplasmin. However, it seems likely that the level of free copper in serum is high as the urinary copper excretion is increased.
      c. treatment with a chelating agent (penicillamine) produces clinical improvement

(ii) Menke's disease
- presents with growth retardation, kinky hair and seizures
- thought to be related to defective copper transport

Zinc

Function
Essential in the metabolism of nucleic acids, in protein synthesis and in many enzyme systems

Clinical deficiency
1. Acquired zinc deficiency
   (i) Acute
   - presents with:
     a. scaling dermatitis
     b. perverted taste and smell
     c. delayed wound healing
     d. diarrhoea
   - associated with:
     a. parenteral nutrition
     b. cirrhosis
     c. coeliac disease
     d. Crohn's disease

   (ii) Chronic
   - this rare form occurs in Iranian and Egyptian men
2. Inborn error of metabolism
   - Acrodermatitis enteropathica (failure to synthesise mucosal zinc-binding protein). Presents in early infancy with diarrhoea, severe skin lesions and alopecia

Toxicity
May result from eating food stored in galvanised containers or use of water from galvanised tanks for home dialysis.

ENTERAL AND PARENTERAL NUTRITION

An adequate supply of energy and nutrients improves resistance to illness and trauma.

Energy requirements are related to:
1. Age
2. Physical activity
3. Body size and composition

Normal adults need approximately 2000–3600 kcal (i.e. Calories) per 24 hours as:
1. Protein (essential; should form at least 10% of total energy)
2. Fat (palatability and fat-soluble vitamins)
3. Carbohydrate (palatability and associated 'roughage')

Vitamins and minerals must also be included in the diet.

Patients who might require enteral or parenteral feeding include those with:

1. Anorexia
   (i) elderly
   (ii) chronically ill
   (iii) psychologically disturbed

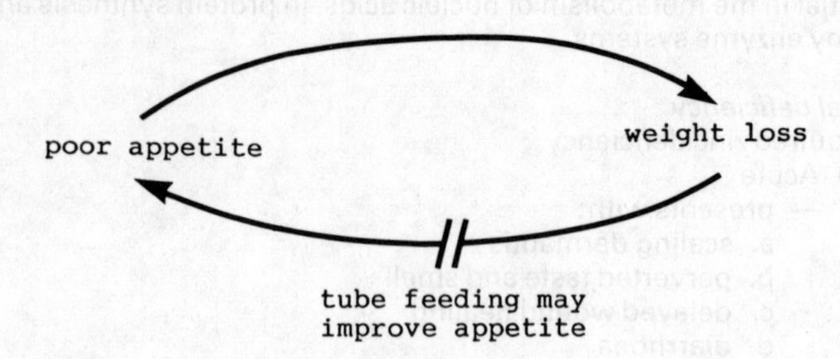

2. Gastrointestinal dysfunction
   (i) fistula
   (ii) inflammatory bowel disease
   (iii) obstruction
   (iv) bowel resection
3. Liver disease
   — affecting metabolism of normal dietary constituents
4. Hypercatabolic states
   — severe injury and infection
   — in these patients:
      a. energy (fat) reserves are mobilised
      b. protein (muscle) reserves are broken down
      c. gluconeogenesis is increased to meet the energy demands of tissue damage

Supply of nutrients
Nutrition should be designed to forestall the use of endogenous stores present in the fat and muscle of normal individuals and to provide a substitute in thin, wasted individuals who have few reserves on which to draw.

N.B. It is easier to prevent loss of body tissues than to regain it later. Each kg gained costs about 8000 kcal of energy and 50 g of nitrogen.

Enteral versus parenteral nutrition
Where the gastrointestinal tract is intact, enteral feeding should be used.

If there is dysfunction in the upper gastrointestinal tract, a feeding jejunostomy can be tried.

If parenteral nutrition cannot be avoided it is best to use a central line and a three-litre bag system.

Energy

If energy is provided as one-third fat to two-thirds glucose, there is a lower incidence of hyperglycaemia, glycosuria and hyperosmolality.

Administration of insulin not only helps to regulate glucose but is also protein-sparing.

An excess of glucose in a severely ill patient will increase oxygen consumption and carbon dioxide production placing a further stress on the respiratory system.

Nitrogen

Nitrogen requirements normally estimated as 1 g per 200 kcal may, during severe illness, be more reliably assessed as 1 g per 150 kcal.

Provided that there is no significant proteinuria or urinary infection, total urine nitrogen loss can be roughly estimated from the 24 hour output of urine urea. Assuming that 80% of total urine nitrogen is urea nitrogen:

$$\text{Approx. 24 h nitrogen excretion (g)} = \frac{\text{mmol urea per 24 h}}{28}$$

Other nitrogen losses from the body amount to around 2 g per 24 hours, so nitrogen replacement can be readily calculated.

Assessment and monitoring

Clinical

1. Six hourly 'stick tests' for glucose and ketones in urine
2. Triceps skinfold thickness } relative changes in fat and muscle mass
3. Mid-arm muscle circumference } relative changes in fat and muscle mass
4. Weight
   — short term reflects fluid balance
   — long term reflects body tissue mass

   N.B Fluid retention (oedema, ascites) or development of excess adipose tissue (e.g. on steroids or excessive glucose administration) may mask the loss of lean body mass.

Laboratory

1. Blood:
   (i) urea, electrolytes
   (ii) glucose (if glycosuria)
   (iii) magnesium
   (iv) phosphate
   (v) zinc

2. Urine:
   (i) 24 hour electrolyte output
   (ii) 24 hour urea excretion

In patients on *prolonged* total parenteral nutrition, further tests may be required:
1. Vitamin B_{12} and folate
2. Serum copper
3. Pre-albumin
   — preferred to albumin as a nutritional marker but value is limited because of its involvement in the acute phase response (see page 137)

Proteins

SERUM PROTEIN

Most serum proteins are synthesised by hepatocytes in the liver, the notable exception being immunoglobulins which are produced by plasma cells.

The serum protein concentration is a balance between synthesis and catabolism and distribution throughout the extravascular pool.

Serum proteins which escape through the capillary walls into the interstitial fluid are returned to the circulation through the lymphatics.

Functions of serum proteins

1. Collective
   - (i) nutritive (act as an aminoacid pool)
   - (ii) osmotic pressure (mainly albumin)
   - (iii) buffering (help to maintain the acid-base balance in the blood)
2. Individual
   - (i) transport (of metals, hormones, drugs etc.)
   - (ii) antibody activity
   - (iii) haem metabolism
   - (iv) coagulation and fibrinolysis
   - (v) protease inhibition
   - (vi) complement activity

Total protein

A normal total protein may mask a low albumin coupled with increased immunoglobulin (as in severe liver disease) whereas dehydration may cause a low total protein to appear normal.

Hyperproteinaemia

1. Apparent increase due to relative water depletion (affects all proteins)
   - (i) dehydration
   - (ii) artefactual (stasis during venepuncture)

2. Increased protein synthesis
   — recognisable hyperproteinaemia usually only occurs with an increase in immunoglobulins (see below)

Hypoproteinaemia

1. Apparent decrease due to relative water excess (affects all proteins)
   (i) overhydration
   (ii) artefactual (sample from 'drip' arm)
2. Unusual loss of protein
   (i) through the kidney
   (ii) through the intestine
   (iii) through the skin (severe burns)
   (iv) severe haemorrhage
   } largely albumin
3. Decreased synthesis
   (i) of albumin
   (ii) of immunoglobulins
   } see below

Serum protein electrophoresis

Serum proteins can be separated by electrophoresis into:

1. albumin
2. alpha-1
3. alpha-2
4. beta
5. gamma

} globulins (2–5)

Albumin

Functions of albumin

1. Maintenance of osmotic pressure
2. Has an important transport function and binds:
   (i) calcium
   (ii) free fatty acids
   (iii) bilirubin
   (iv) drugs
   (v) a variety of other substances

Hyperalbuminaemia

Always caused by relative water depletion

Hypoalbuminaemia

— *Causes*:

1. Decreased synthesis
   (i) severe chronic liver disease
   (ii) deficiency of aminoacid supply (malnutrition, malabsorption)
   (iii) genetic defect (analbuminaemia)—very rare

2. External loss
   (i) renal (nephrotic syndrome)
   (ii) gastrointestinal (protein-losing enteropathy)
   (iii) through injured skin (severe burns, exfoliative dermatitis)
   (iv) frequent removal of ascitic fluid
3. Increased endogenous catabolism
   (i) malignant tumours
   (ii) infection
   (iii) trauma
4. Redistribution
   — in many pathological states there is a shift of albumin towards the extravascular space, probably due to increased capillary permeability and an impaired venous return

N.B.
— serum albumin levels below 30 g/l usually indicate serious organic disease
— values between 20–30 g/l are most commonly found in malignant conditions or liver disease
— levels below 20 g/l are more often associated with renal or enteral protein-losing states

Oedema and serous effusions (see also page 9)
Oedema is almost always present when the serum albumin concentration falls below 20 g/l.

Oedema fluid or *transudate* is produced through capillaries with almost normal permeability and has a low protein content (often < 10 g/l).

Exudate is produced through capillaries with increased permeability and the protein concentration reflects the degree of vascular damage (usually > 30 g/l).

Globulins

Immunoglobulins
Most of the immunoglobulins are found in the gamma globulin fraction.

Functions
IgG Protection of tissue spaces
IgA Protection of skin and mucosal surfaces
IgM Protection of the bloodstream
IgE Involvement in local anaphylactic reactions

Clinical conditions in which immunoglobulin assay may be helpful
1. Unexplained high ESR
2. Suspected hypogammaglobulinaemia (primary or secondary) e.g. patients with recurrent infections

3. Pyrexia of unknown origin
4. Liver disease
5. Neonatal infection (raised IgM suggests intrauterine infection)

Hypergammaglobulinaemia

1. Diffuse (polyclonal)—commonly found with: (see below)
   (i) acute and chronic infections
   (ii) chronic liver disease
   (iii) autoimmune disease
   (iv) granulomatous states
2. Discrete (monoclonal) (see below)
   (i) benign
   (ii) malignant

Polyclonal hypergammaglobulinaemia
— *Causes*:

1. Mainly IgG
   (i) systemic lupus erythematosus
   (ii) chronic active hepatitis
2. Mainly IgA
   (i) Laennec's cirrhosis
   (ii) Crohn's disease
   (iii) coeliac disease
   (iv) tuberculosis
   (v) sarcoidosis
   (vi) dermatomyositis
3. Mainly IgM
   (i) primary biliary cirrhosis
   (ii) perinatal infection
   (iii) tropical diseases (various)
4. Equal % increase in IgG, IgA and IgM
   (i) most infections (long standing)
   (ii) late sarcoidosis
   (iii) late liver disease

Elevated IgE (diffuse)
— this is never sufficient to cause hypergammaglobulinaemia

— *Causes*:
1. atopy
2. asthma
3. allergic rhinitis
4. parasitic infestation

Monoclonal hypergammaglobulinaemia (paraproteinaemia)
— *Causes*:
1. Malignant
   (i) multiple myeloma

(ii) Waldenström's macroglobulinaemia
(iii) soft tissue plasmacytoma
(iv) occasionally in lymphomas and chronic lymphocytic leukaemia

2. Benign
   (i) 'benign monoclonal gammopathy'
   — a paraprotein can only be considered benign when it has been followed for at least 5 years and there has been no increase in concentration
   — *N.B.* the prevalence of paraproteins in a healthy population increases with age:

$<$ 60 years	0.2%
61–80	1.6%
81–90	11.7%
$>$ 90 years	19.0%

   (ii) monoclonal antibody
   — often transient

— *Features suggesting malignancy in paraproteinaemia*:
1. Characteristic bone marrow findings
2. Characteristic X-ray findings
3. Paraprotein in excess of:
   20 g/l for IgG
   10 g/l for IgA
   10 g/l for IgM
4. IgD and IgE paraproteinaemia is always malignant
5. Bence-Jones proteinuria in excess of 1 g/l
6. Increasing paraprotein level
7. Immunosuppression (i.e. decreased levels of the normal Ig's)

Multiple myeloma
The diagnosis of multiple myeloma (as opposed to 'paraproteinaemia') is a clinical one and requires the presence of at least *two* of the following diagnostic criteria:
1. a monoclonal protein in serum (i.e. a paraprotein) or urine (i.e. Bence-Jones protein)
2. characteristic X-ray appearance
3. characteristic bone marrow findings

— *Common presentations include*:
1. back pain
2. raised ESR
3. anaemia
4. pneumonia
5. hypercalcaemia
6. pathological fractures
7. impaired renal function

— *Biochemical investigation includes*:
1. estimation of total protein in serum and urine
2. identification and quantification of any serum paraprotein
3. quantification of the unaffected immunoglobulin classes (to assess immunosuppression)
4. detection of any free light chains in urine (Bence-Jones proteinuria)
5. quantification of Bence-Jones protein (if there is Bence-Jones proteinuria but no serum paraprotein)
6. estimation of serum albumin, urea, calcium and urate

Incidence of different classes of paraprotein in multiple myeloma:

IgG	53%
IgA	22%
Bence-Jones only*	20%
IgD	1.5%
IgM	0.5%

* Bence-Jones only myeloma may be missed if urine is not examined

Very rarely, 'non-secretory myeloma' occurs in which neither serum paraprotein nor Bence-Jones protein is found.

IgM paraproteinaemia
1. Waldenström's macroglobulinaemia (50%)
   — occurs more commonly in males than females
   — anaemia and lymphadenopathy are common, skeletal changes rare
   — may present as 'hyperviscosity syndrome' (*N.B.* large molecular size of IgM)
2. Malignant lymphoma (25%)
3. Benign monoclonal gammopathy (10%)
4. Myeloma—this is a *rare* cause

Hypogammaglobulinaemia
1. Transient (IgG)
   (i) premature babies (at birth)
   (ii) normal babies (from 3–6 months)
2. Primary
   (i) IgA deficiency (1 in 500 people)
   (ii) other rare defects of the immune system (usually hereditary) with absence of one or more immunoglobulins, e.g.
   — Bruton's sex-linked agammaglobulinaemia
   — Swiss-type agammaglobulinaemia
3. Secondary
   — to acquired disease or to therapy (approx. 4% of hospital patients)

(i) lymphatic malignancies
  a. chronic lymphocytic leukaemia
  b. malignant lymphomas
  c. multiple myeloma
  d. Waldenström's macroglobulinaemia
(ii) protein-losing states
  a. nephrotic syndrome
  b. protein-losing enteropathies
  c. exfoliative dermatitis
(iii) uraemia
(iv) immunosuppressive therapy
  a. drugs (steroids, cyclophosphamide, azathioprine)
  b. irradiation

Other globulins

Alpha-1-antitrypsin
Alpha-1-antitrypsin is responsible for 90% of the protease inhibition present in serum. Deficiency of the protein can result in damage due to unchecked protease activity. A number of genetic variants of this protein have been identified which are associated with decreased protease activity.

Homozygous alpha-1-antitrypsin deficiency is associated with:
1. emphysema
2. cirrhosis

Transferrin
— see page 117

The acute phase reaction
This is a non-specific change in the concentration of several proteins following tissue damage e.g. trauma, infection, malignancy. It is often associated with a raised ESR (as is monoclonal or polyclonal hypergammaglobulinaemia)
1. Proteins whose concentration increases:
   (i) alpha-1 and alpha-2 globulins (alpha-1-antitrypsin, alpha-1-antichymotrypsin, orosomucoid, haptoglobin)
   (ii) fibrinogen
   (iii) C-reactive protein
2. Proteins whose concentration decreases:
   (i) albumin
   (ii) pre-albumin
   (iii) transferrin (a beta globulin)

N.B. the major finding on serum protein electrophoresis is an increase in alpha-1 and alpha-2 globulins.

URINE PROTEIN

Proteinuria of more than 0.15 g/24 hours always indicates disease.

Classification of proteinuria

1. Pre-renal proteinuria
   — low molecular weight protein loss with normal renal function
   (i) Bence-Jones protein
   (ii) haemoglobin
   (iii) myoglobin
   (iv) low molecular weight acute phase proteins
2. Renal proteinuria
   (i) Glomerular
   — increased glomerular permeability
   a. orthostatic (postural)
   b. glomerular disease

   N.B. massive proteinuria is *always* of glomerular origin

   (ii) Tubular
   — damage to renal tubules
   — protein loss is less than 1 g/24 hours
3. Post-renal proteinuria
   — damage to the ureters or bladder (e.g. by calculi, infection or malignancy) may produce low grade proteinuria

Nephrotic syndrome

— see page 27

CSF PROTEIN

Causes of increased CSF protein concentration:

1. Increased capillary permeability
   — allowing passage of an excess of plasma protein into the CSF
   (i) bacterial or viral meningitis
   (ii) intracranial neoplasm
   (iii) cerebrovascular disease
2. Mechanical obstruction
   — progressive equilibration of static CSF with plasma
3. Excessive synthesis of immunoglobulins within the CNS
   (i) tuberculous meningitis
   (ii) neurosyphilis
   (iii) multiple sclerosis

The demonstration of increased immunological activity within the CNS, (either through the presence of oligoclonal banding of CSF IgG on polyacrylamide electrophoresis, or by measurement of the IgG: albumin ratio in the CSF and serum) is most often requested in suspected multiple sclerosis but is not diagnostic.

N.B. IgG may be synthesised in the CNS without significantly increasing the total CSF protein

4. Contamination with blood
   (i) subarachnoid haemorrhage
   (ii) traumatic lumbar puncture

N.B. lumbar CSF normally has a higher protein content than that obtained from the ventricles.

Biochemical markers of malignancy

Hypercalcaemia (see page 90)
Hypercalcaemia may be a marker of malignancy

Increase in alkaline phosphatase
1. Malignancy of bone—when chiefly osteoblastic
   (i) primary
   (ii) secondary
2. Malignancy of liver
   (i) primary
   (ii) secondary
3. Hyperparathyroidism
   — may be caused by parathyroid carcinoma
4. Ectopic production of alk. phos. by a tumour
   — especially carcinoma of the bronchus
   — the isoenzyme produced often has atypical mobility on electrophoresis. The so-called 'Regan' isoenzyme is very similar to placental alk. phos.

Increase in acid phosphatase (see page 163)
— carcinoma of the prostate
   (i) without metastases (25%)
   (ii) with skeletal metastases (up to 90%)
   (iii) with only soft tissue metastases (less frequent and lower levels than with bone involvement)
— unreliable in early disease
— unnecessary in late disease
— unsatisfactory for monitoring
— may not be produced at all by very poorly differentiated tumours

Production of monoclonal immunoglobulins (paraproteins)
— see page 134

Hormones

1. Hypersecretion of normal hormones in the absence of normal regulating mechanisms (see chapter VIII)
   — sometimes caused by carcinomas of endocrine glands
2. Ectopic secretion of hormones by tumours of non-endocrine tissue
   — carcinoma of the bronchus is by far the commonest tumour to produce ectopic hormones especially:
   (i) antidiuretic hormone (see page 7)
   (ii) ACTH (see page 80)
   (iii) parathyroid hormone
3. Inappropriate hormone secretion from two or more endocrine glands (see 'multiple endocrine adenopathy' page 87)

Onco-fetal markers

1. Carcinoembryonic antigen (CEA)
   (i) found in fetal intestine, liver and pancreas up to 6 months gestation
   (ii) sometimes found in malignant tissue especially:
      a. colon
      b. liver
      c. pancreas

   Lack of specificity reduces the value of CEA in diagnosis but serum levels may be used to monitor response to treatment and then for follow-up.
2. Alpha-fetoprotein (AFP)
   (i) synthesised by fetal liver
      — maximal about 13 weeks
      — synthesis ceases a few weeks after delivery
   (ii) synthesised by malignant tissue of
      a. primary hepatocellular tumours
         — not all produce AFP and the frequency of production seems to vary from country to country
         — the incidence of elevated serum AFP also depends on the detection limits of the assay employed
         — other non-malignant diseases may produce mild elevations of AFP e.g. hepatitis, alcoholic cirrhosis, ulcerative colitis
      b. teratomas of testis or ovary
         — good correlation between size of tumour and AFP level so useful for monitoring therapy

3. Chorionic gonadotrophin (beta subunit i.e. beta-HCG)
   (i) found in serum during pregnancy
   (ii) found in tumours
      a. male
         — teratoma of testis
      b. female
         — choriocarcinoma
         — hydatidiform mole

   Of value in the detection and monitoring of these tumours

   Ectopic secretion may also occur from:
   — adenocarcinoma of stomach, ovary or pancreas
   — hepatoma

Lipids

PHYSIOLOGY

Lipids are transported in plasma as pseudomicelles called *lipoproteins* and consisting of:

1. a centre of non-polar lipids (triglycerides and cholesterol ester)
2. a surface layer containing:
   - (i) more polar lipids (free cholesterol and phospholipid)
   - (ii) proteins (apolipoproteins)

The different lipoproteins are shown in Table 7.

Lipoprotein metabolism

Fig. 17 shows the transport of lipids by the various lipoproteins. The arrows do not imply the interconversion of lipoproteins except where the arrow contains a solid component representing apolipoprotein B (apo B). In other cases they simply show the transfer of lipid.

Table 7 Plasma lipoproteins

Class	Electrophoretic mobility	Diameter (nm)	Major lipids (% of total mass)
Chylomicra	Origin	100–1000	Glyceride (85%)
Very low density lipoprotein (VLDL)	Pre-beta	30–80	Glyceride (50%)
Intermediate density lipoprotein (IDL)	Beta	25–30	Glyceride (35%) Chol. ester (25%)
Low density lipoprotein (LDL)	Beta	20–25	Chol. ester (40%)
High density lipoprotein (HDL)	Alpha	5–10	Phospholipid (25%) Chol. ester (15%)

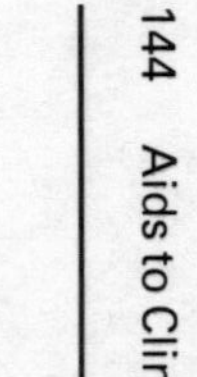

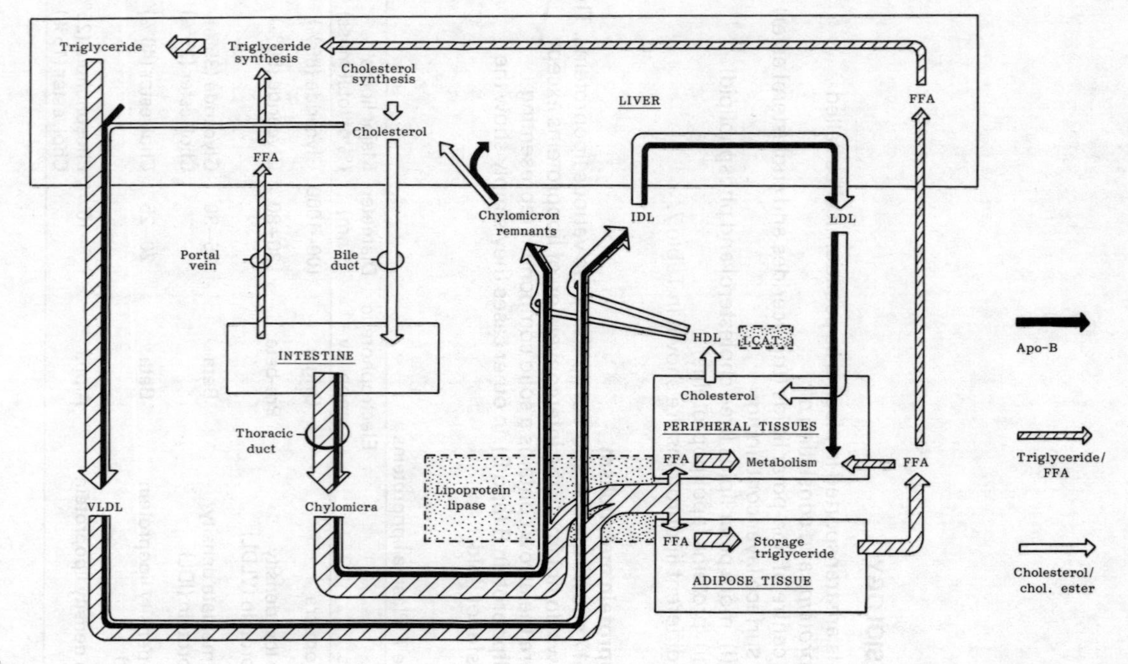
Triglyceride
Triglyceride
synthesis
Cholesterol
synthesis
Cholesterol
FFA
LIVER
FFA
IDL
LDL
Chylomicron
remnants
Portal
vein
Bile
duct
INTESTINE
HDL
LCAT
Cholesterol
PERIPHERAL TISSUES
Thoracic
duct
FFA
Metabolism
FFA
Lipoprotein
lipase
VLDL
Chylomicra
FFA
Storage
triglyceride
ADIPOSE TISSUE
Apo–B
Triglyceride/
FFA
Cholesterol/
chol. ester

Fig. 17

Triglyceride transport

1. The *triglyceride-rich lipoproteins* are:
   (i) Chylomicra
   — these transport triglyceride derived from dietary long chain fatty acids.
   (ii) VLDL
   — these transport triglyceride synthesised in the liver.
2. Lipoprotein lipase (LPL)
   — this enzyme found in capillary walls hydrolyses triglyceride (of chylomicra and VLDL) to free fatty acids (FFA) and glycerol
   — this leads to conversion of:
   chylomicra → chylomicron remnants
   VLDL → IDL
   — the FFA released are either reconverted to triglyceride for storage in adipose tissue or are metabolised to provide energy
3. Hormone-sensitive lipase
   — in the non-fed state, this enzyme hydrolyses adipose tissue triglyceride, releasing FFA which may then be utilised by other tissues or converted into VLDL triglyceride in the liver
4. Dietary short chain fatty acids pass directly to the liver via the portal vein

Cholesterol transport

1. Dietary cholesterol is incorporated into chylomicra which are in turn converted into chylomicron remnants and the latter are taken up into the liver
2. Hepatic cholesterol may be:
   (i) excreted in the bile (either unchanged or after conversion to bile salts)
   or
   (ii) exported as a component of VLDL
3. VLDL are converted to IDL which are in turn converted to LDL, the latter conversion probably occurring mainly in the liver
4. LDL are taken up by peripheral cells by means of cell surface receptors which recognise the apo B in the LDL (other apo B containing lipoproteins are probably too large to gain access to the interstitial fluid and so are not taken up)
5. The uptake of LDL by peripheral cells:
   (i) switches off their own cholesterol synthesis
   (ii) reduces the production of the surface receptors
6. HDL
   — excess free cholesterol is picked up from cells by HDL
   — within HDL, cholesterol is converted to esters by the enzyme lecithin-cholesterol acyl transferase (LCAT)
   — much of this cholesterol ester appears to transfer to chylomicron remnants and IDL and so is returned to the liver for re-use or excretion

Apolipoproteins

1. Apo A-I and apo A-II
   — major components of HDL (apo A-I being an activator of LCAT)
2. Apo B
   — structural protein for chylomicra, chylomicron remnants, VLDL, IDL and LDL
3. Apo C-I, apo C-II and apo C-III
   — normally stored in HDL between meals but have a higher affinity for triglyceride-rich lipoproteins and so transfer to these when they appear in plasma after meals.
   — apo C-II is an activator of LPL and as the triglyceride is removed from the triglyceride-rich lipoproteins by LPL, the apo C's return to HDL

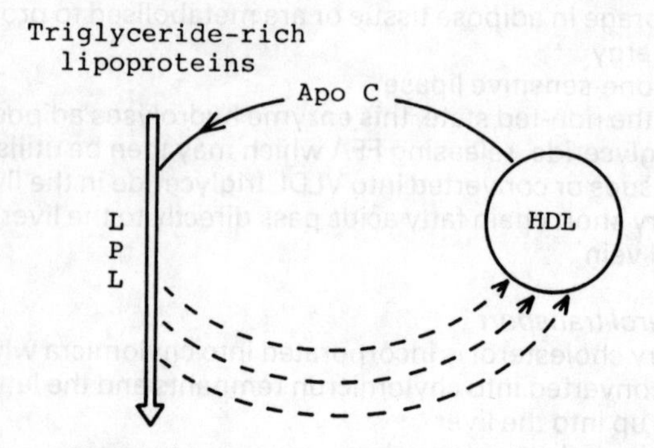

4. Apo E-I, apo E-II, apo E-III and apo E-IV
   — apo E enters plasma with nascent HDL produced in the liver
   — appears to transfer with cholesterol ester from HDL to triglyceride-rich lipoprotein remnants (thus by the time they have become remnants, the triglyceride-rich lipoproteins have lost most of their triglyceride and gained cholesterol ester and apo E)
   — seems to be important in the hepatic recognition of these remnants for hepatic uptake

Effects of diet on serum lipids

LDL and VLDL levels are lowered by a diet:

1. low in saturated fat
2. low in cholesterol
3. with a calorie content to overcome obesity

Effects of exercise on serum lipids
Regular exercise:
1. lowers LDL levels
2. raises HDL levels
3. probably speeds up the removal of triglyceride-rich lipoproteins

ATHEROMA AND CORONARY ARTERY DISEASE

The major reason for clinical interest in lipids is their role in the development of atheroma especially in the coronary arteries.

However, it is important to note that other factors are also important in the aetiology of atheroma:

Risk factors for coronary artery disease

The major risk factors are:
1. Hypertension
2. Lipid risk factors (see below)
3. Glucose intolerance
4. Cigarette smoking

Lipid risk factors (see 'laboratory assessment' below)
1. High total cholesterol
   — the most well defined risk factor
   — normally LDL is the greatest contributor to total serum cholesterol and it is now high LDL-cholesterol levels which are believed to be important in the majority of cases
2. Low HDL-cholesterol
3. Other less well defined factors, e.g.
   (i) high concentrations or delayed metabolism of remnant particles (from triglyceride-rich lipoproteins)
      — these particles are atherogenic, as demonstrated by 'remnant hyperlipidaemia' (see page 152), but some believe that they may also be of prime importance in many other cases of atheroma
      — the protective effect of exercise may be partly related to faster metabolism of these particles
   (ii) high triglyceride levels
      — now believed by many to be of little direct importance (though often associated with low HDL-cholesterol levels)

LABORATORY ASSESSMENT OF LIPID LEVELS

The following investigations may be performed on blood taken after a 12-14 hour fast:
1. Appearance of serum after standing overnight at 4°C:
   (i) 'cream' layer at the top
      — demonstrates the presence of chylomicra

(ii) turbid sample
— increased VLDL concentration
— *N.B.* LDL and HDL are too small to cause light scattering so they never cause turbidity in samples

2. Total cholesterol
3. Triglyceride
4. HDL-cholesterol
   — the apo B containing lipoproteins are precipitated, then cholesterol is measured in the supernatant
   — not routinely available in many laboratories
5. Lipoprotein electrophoresis
   — rarely performed as a routine investigation now

N.B.
(i) total cholesterol and HDL-cholesterol are only affected to a small extent by fasting and so can be measured on non-fasting samples (though difficulties with the precipitation step of HDL cholesterol measurement may occur with some samples)
(ii) lipid levels tend to be lowered for up to 3 months after a myocardial infarction, major surgery etc.

HYPERLIPIDAEMIAS

Hyperlipidaemia is the presence of serum lipid levels above a given arbitrarily defined cut-off concentration (often the 95th percentile).

In the normal population, serum cholesterol and triglyceride concentrations both increase with age and are different in men and women, so ideally age/sex related reference intervals should be employed (but unfortunately, this is often not done).

Hyperlipidaemia can simply be described as being:
Hypercholesterolaemia
Hypertriglyceridaemia
Mixed hyperlipidaemia (both cholesterol & triglyceride ↑)

World Health Organisation (W.H.O.) classification:
— separates hyperlipidaemia on the basis of the *pattern of lipoprotein elevation(s)*
— it gives no more than clues about the aetiology of the hyperlipidaemia
— it can be used to describe the pattern of lipoprotein elevation in both primary and secondary hyperlipidaemias (see below)
— where possible it is better to use a classification based on aetiology (see 'genetic classification' below)
— the W.H.O. classification is shown in Table 8 along with the genetic classification equivalents

Table 8 Primary hyperlipidaemias

SERUM				W.H.O. CLASSIFICATION	POSSIBLE CAUSES (Genetic classification)	DEFECT
APPEARANCE	CHOL.	TRIG.	LIPOPROTEINS INCREASED			
Clear with cream layer	N (or $\uparrow$)	$\uparrow\uparrow\uparrow$	Chylomicra	type I	Familial lipoprotein lipase deficiency	Lipoprotein lipase deficiency
Clear	$\uparrow\uparrow\uparrow$	N	LDL	type IIa	Familial hypercholesterolaemia Familial combined hyperlipidaemia Polygenic	LDL receptor defect $\uparrow$ hepatic apo B production
Turbid	$\uparrow\uparrow$	$\uparrow\uparrow$	LDL & VLDL	type IIb	Familial combined hyperlipidaemia Bigenic or Polygenic	$\uparrow$ hepatic apo B production
Turbid + cream layer	$\uparrow\uparrow\uparrow$	$\uparrow\uparrow\uparrow$	Remnants	type III	Remnant hyperlipidaemia	Apo E-III deficiency PLUS familial combined hyperlipidaemia*
Turbid	N or $\uparrow$	$\uparrow\uparrow$	VLDL	type IV	Familial combined hyperlipidaemia Familial hypertriglyceridaemia Polygenic	$\uparrow$ hepatic apo B production $\uparrow$ hepatic triglyceride production
Turbid with cream layer	N or $\uparrow$	$\uparrow\uparrow\uparrow$	VLDL & chylomicra	type V	Familial hypertriglyceridaemia (homozygous)**	$\uparrow\uparrow$ hepatic triglyceride production

* or apo E-III deficiency plus a secondary hyperlipidaemia
** or heterozygote plus a secondary hyperlipidaemia

Genetic classification

1. Primary hyperlipidaemias
   (i) Monogenic
      a. familial hypercholesterolaemia
      b. familial combined hyperlipidaemia
      c. familial hypertriglyceridaemia
      d. familial lipoprotein lipase deficiency
      e. remnant hyperlipidaemia
   (ii) Polygenic
2. Secondary hyperlipidaemias

If two monogenic primary hyperlipidaemias occur simultaneously, this is often called a 'bigenic' hyperlipidaemia. Stricly speaking, remnant hyperlıpidaemia is usually a bigenic disorder (see page 152).

N.B. Diet and exercise affect the serum lipid levels in *all* of the above

Table 9. Inheritance and prevalence of the primary hyperlipidaemias

	Probable pattern of inheritance	Prevalence in the general population
Familial hypercholesterolaemia	autosomal dominant	1 in 500
Familial combined hyperlipidaemia	autosomal dominant	1 in 300
Familial hypertriglyceridaemia	autosomal dominant	1 in 500
Remnant hyperlipidaemia	complex (see text)	rare
Familial lipoprotein lipase deficiency	autosomal recessive	rare

PRIMARY HYPERLIPIDAEMIAS

Monogenic hyperlipidaemias

Familial hypercholesterolaemia (FH)
FH shows a gene dosage effect i.e. the rare homozygotes are more severely affected than heterozygotes.

— *Aetiology*:

Cell surface receptors for LDL deficient or defective → high LDL levels in the blood → deposition of cholesterol in tissues

— *Clinical features*:
1. Premature corneal arcus
2. Tendon xanthomas
3. Xanthelasmas
4. High incidence of myocardial infarction
   — 85% of heterozygotes have a myocardial infarction by 60 years of age
   — even though FH has a prevalence of only 1 in 500 in the general population, it is present in 1 in 20 patients suffering a myocardial infarction

— *Serum lipids*:

Serum cholesterol:	7–16 mmol/l	(heterozygotes)
	16–24 mmol/l	(homozygotes)

W.H.O. type: always IIa (unless combined with another hyperlipidaemia)

Familial combined hyperlipidaemia
— *Aetiology*:
Probably hepatic overproduction of the apo B component of VLDL

— *Clinical features*:
1. There may be corneal arcus or xanthelasmas but *no tendon xanthomas*
2. Increased incidence of coronary artery disease

— *Serum lipids*:
The pattern depends on the rate of conversion of VLDL to LDL which in turn depends on polygenic factors, body weight, diet, exercise etc. Affected members of a family often have different lipid patterns and even one individual may have a varying pattern from time to time.

W.H.O. type:

IV	(increased VLDL)	approx. one third of patients show each of these patterns
IIa	(increased LDL)	
IIb	(increased VLDL & LDL)	
III		may be found on rare occasions (see below)
V		

Familial hypertriglyceridaemia
— *Aetiology*:
Probably hepatic overproduction of the triglyceride component of VLDL

— *Clinical features*:
Often unremarkable unless there is severe chylomicronaemia (see below)

There is probably no increased incidence of coronary artery disease

— *Serum lipids*:
Hypertriglyceridaemia

W.H.O. type:
usually type IV (increased VLDL)
occasionally type V* (increased VLDL & chylomicra)

Remnant hyperlipidaemia
— *Aetiology*:
Remnant hyperlipidaemia requires *both*:
1. Apo E-III deficiency (sometimes called 'familial dysbetalipoproteinaemia', though this term has been used synonymously with remnant hyperlipidaemia)
   — autosomal recessive inheritance
   — prevalence: 1 in 100 of general population
   — apo E-III deficiency → impaired hepatic uptake of remnants
   — *no* hyperlipidaemia on its own

and

2. Another hyperlipidaemia which causes an increased production rate of remnants
   — usually familial combined hyperlipidaemia
   — may be a secondary hyperlipidaemia

— *Clinical features*:
1. Palmar crease xanthomas
2. Tuberous xanthomas
3. Increased incidence of *peripheral vascular disease* and coronary artery disease

— *Serum lipids*:
Mixed hyperlipidaemia
W.H.O. type: III (increased remnants ± chylomicra)
Lipoprotein electrophoresis shows a band from the beta to the prebeta region ('broad beta') caused by the remnants.

Familial lipoprotein lipase deficiency
— *Aetiology*:
Deficiency of lipoprotein lipase

— *Clinical features*:
Those of chylomicronaemia (see below)
No increased incidence of coronary artery disease

* (i) homozygotes for this condition *or*
(ii) heterozygotes who also have a secondary hypertriglyceridaemia

— *Serum lipids*:
Hypertriglyceridaemia
W.H.O. type: I (presence of chylomicra in fasting sample)

Polygenic hyperlipidaemias
1. Multiple gene effect i.e. no single abnormal gene
2. Hyperlipidaemic parents are more likely to have hyperlipidaemic offspring but the offspring do not divide clearly into those affected and those not affected as they would with a monogenic hyperlipidaemia
3. Prevalence not clearly defined as there is no single genetically defined defect—they simply represent the upper extreme of the normal distribution of lipid levels—their definition is arbitrary

SECONDARY HYPERLIPIDAEMIA

Common causes
1. Predominant hypercholesterolaemia
   — cholestasis
   — hypothyroidism
   — nephrotic syndrome*
2. Predominant hypertriglyceridaemia
   — diabetes
   — alcoholism
   — chronic renal failure
   — nephrotic syndrome*

* Nephrotic syndrome may give either hypercholesterolaemia or hypertriglyceridaemia

CLINICAL APPROACH TO DIAGNOSIS

Hypercholesterolaemia with normal triglyceride
1. Exclude secondary causes
2. Exclude elevated HDL-cholesterol as a cause (if the assay is available)
3. Consider monogenic primary causes:
   (i) familial hypercholesterolaemia:
      — tendon xanthomas
      — serum cholesterol usually > 9 mmol/l
      — family study
   (ii) familial combined hyperlipidaemia
      — no tendon xanthomas
      — serum cholesterol usually < 9 mmol/l
      — family study
4. Having excluded the above: probably polygenic

Hypertriglyceridaemia + or − hypercholesterolaemia

1. Exclude secondary causes
2. Consider monogenic primary causes:
   - (i) familial combined hyperlipidaemia
   - (ii) familial hypertriglyceridaemia

   } family studies are the only routine method available for distinguishing between these

   - (iii) familial LPL deficiency
     — chylomicronaemia, especially in a child, and in the absence of increased VLDL, is an indication for measuring LPL in a blood sample taken after the injection of heparin (the latter frees LPL from capillaries into the circulation where it can be measured)
   - (iv) remnant hyperlipidaemia
     — palmar crease xanthomas
     — tuberous xanthomas
     — total cholesterol > 9 mmol/l and total cholesterol = approx. 2 × triglyceride

N.B. Since some monogenic hyperlipidaemias are fairly common, it is not unusual for a patient to have two different abnormal genes (bigenic disorder):
e.g. familial hypertriglyceridaemia plus familial hypercholesterolaemia → W.H.O. type IIb pattern

3. Having excluded the above: probably polygenic

N.B. Chylomicronaemia can occur in any of the conditions listed in 1. and 2. above

Clinical features of chylomicronaemia

1. Eruptive xanthomas
2. Abdominal pain
3. Lipaemia retinalis

TREATMENT

When to treat hyperlipidaemia

Always treat:

1. Triglyceride > 10 mmol/l (risk of pancreatitis)
2. Cholesterol > 9 mmol/l

Factors suggesting treatment at lower lipid levels:

1. Young age
2. Bad family history
3. Associated risk factors

Mode of treatment

1. Look for secondary hyperlipidaemia and if found treat the primary disease
2. Diet
3. Exercise
4. Drugs (when all other measures fail)
   — Cholestyramine is the most well proven and safest drug for the treatment of hypercholesterolaemia (when caused by increased serum LDL concentration). It is not absorbed from the gut but acts by binding bile salts and interfering with their reabsorption in the small bowel. Since bile salts are synthesised from cholesterol in the liver, their loss leads to increased hepatic uptake of LDL, so lowering the serum LDL concentration.

Enzymes

INTRODUCTION

Most enzymes remain largely within the cells in which they are produced. Exceptions are:

1. Enzymes involved in coagulation
2. Enzymes secreted into the gastrointestinal tract

Consequently, serum enzyme levels are low and result from normal cell breakdown, the enzymes becoming inactive within a few days of entering the circulation.

Only a few enzymes are measured for routine clinical use and those to be considered here will be abbreviated as follows:

Aspartate aminotransferase	AST
Alanine aminotransferase	ALT
Lactate dehydrogenase	LDH
Creatine kinase	CK
α –Amylase	Amylase
Alkaline phosphatase	Alk. phos.
Acid phosphatase	Acid phos.
γ –Glutamyl transferase	γGT
5′–Nucleotidase	5NT

Mechanism of enzyme changes

The serum level of each enzyme normally reflects the balance between production (about which much is known) and disappearance (about which next to nothing is known); in disease, the level also depends on the amount of enzyme in the affected tissue and the extent of the damage.

Distribution of enzymes throughout the body

	LDH	AST	ALT	CK	Alk. Phos.	Acid Phos.
Liver	++++	++++	++++	+	+++	+
Cardiac Muscle	++++	++++	++	++	+	
Skeletal Muscle	++++	+++	++	++++	+	
Bone					++++	+
Prostate					++	++++

Production
Serum enzyme levels may be elevated because of:
1. Increased synthesis by existing cells
   e.g. γGT induced by alcohol or other microsomal enzyme-inducing agents (like phenytoin or phenobarbitone)
2. Increase in number of synthesising cells
   e.g. acid phos. from prostatic metastases
3. Leakage from injured or dying cells
   Rate of release of enzymes from damaged cells depends on:
   (i) Location of enzyme within the cell
      a. cytoplasmic
      b. mitochondrial
      c. microsomal
   (ii) Cell/serum concentration gradient
   (iii) Molecular size
   (iv) Membrane permeability which may be increased by:
      a. restriction of oxygen supply to the cell
      b. depletion of circulating glucose
      c. depletion of cell energy production (drugs or chemicals)
      d. bacterial toxins or viruses
      e. genetic defect of membrane structure (e.g. muscular dystrophy)

Serum enzyme levels may be low:
1. When too many synthesising cells have been destroyed
2. In congenital deficiency

Disappearance
Enzymes disappear from the circulation at different rates:

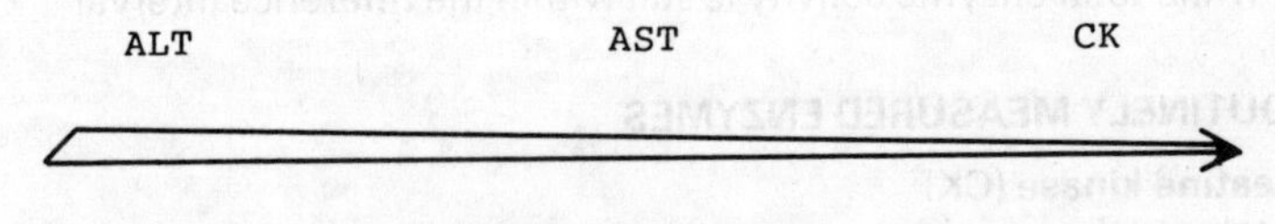

Of the routinely measured enzymes, *only amylase has renal excretion*
(*N.B.* serum amylase rises in renal failure)

Increases unrelated to specific pathology
These may be related to:
1. Age and sex
2. Pregnancy and lactation
3. Enzyme induction by drugs
4. Haemolysis of the sample

Isoenzymes
Isoenzymes are multiple forms of an enzyme catalyzing the same reaction, but occurring in different physico-chemical forms.
Strictly, the term should only be applied to the multiple forms of an enzyme which are the result of synthesis controlled by two or more genes. In many cases, the enzyme in question is composed of sub-units, each of which exists in two or more varieties and each variety is the product of a different gene. Thus if an enzyme has two varieties of sub-unit, and it exists as a dimer (2 sub-units per molecule), it will have 3 isoenzymes:

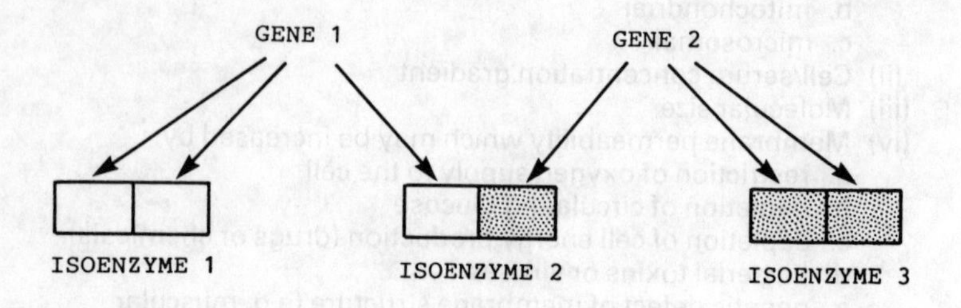

For any given enzyme, the proportion of each isoenzyme is often different in different tissues.

Diagnostic advantages:
The isoenzyme distribution may implicate a particular tissue or organ source:
1. when there is an increase in total enzyme activity

or

2. while total enzyme activity is still within the reference interval

ROUTINELY MEASURED ENZYMES

Creatine kinase (CK)
Catalyses the reaction:

$$\text{Creatine} + \text{ATP} \rightleftharpoons \text{Creatine phosphate} + \text{ADP}$$

CK is found in *skeletal* and *heart muscle* and to a lesser extent in brain.

Causes of increased serum total CK:
1. Myocardial infarction (see page 165)
2. Muscular dystrophies (see page 166)
3. Polymyositis (see page 167)
4. Hypothyroidism

5. Muscle trauma (including intramuscular injections, surgical procedures, severe exercise, epileptic seizures)
6. Hypoxic muscle damage
7. Alcoholic myopathy
8. Coma from hypnotic overdose
9. Pulmonary infarction

CK isoenzymes
There are 2 types of sub-unit: 'M' (muscle) and 'B' (brain)
CK exists as a dimer (2 sub-units) so there are 3 isoenzymes:
CKMM — preponderant form in skeletal and cardiac muscle
CKBB — preponderant form in brain
CKMB — present in cardiac muscle (minor component)
— normally undetectable in serum

Clinical application of CKMB measurement:
Useful for the investigation of myocardial infarction especially:
1. to differentiate it from other causes of a raised total CK, e.g.
   — hypoxia (as often occurs with a cardiac arrest)
   — recent trauma (including surgery)
2. in the presence of a normal total CK
3. to assess the extent of myocardial infarction (by means of serial measurements of CKMB), though this is not generally performed as a routine

Lactate dehydrogenase (LDH)
Catalyses the reaction:

$$\text{Lactate} + \text{NAD}^+ \rightleftharpoons \text{Pyruvate} + \text{NADH} + \text{H}^+$$

Because of its involvement in glucose metabolism LDH is widely distributed, but the richest tissues are *cardiac* and *skeletal muscle, liver, kidney* and *red blood cells.*

Causes of increased serum total LDH:
1. Megaloblastic anaemia (see page 167)
2. Widespread carcinomatosis, especially hepatic metastases
3. Shock/hypoxia (→ liver, muscle damage)
4. Hepatocellular damage (but other enzymes are more useful)
5. Myocardial infarction
6. Pulmonary infarction
7. Haemolytic conditions
8. Leukaemias
9. Skeletal muscle disease e.g. muscular dystrophy

LDH isoenzymes
There are 2 types of sub-unit: 'H' (heart) and 'M' (muscle)
LDH exists as a tetramer (4 sub-units) so there are 5 isoenzymes:

LD_1 (i.e. H_4)	— fastest electrophoretic mobility
LD_2 (i.e. H_3M)	↓
LD_3 (i.e. H_2M_2)	↓
LD_4 (i.e. HM_3)	↓
LD_5 (i.e. M_4)	— slowest electrophoretic mobility

In normal serum LD_1, LD_2 and LD_3 are readily detected by electrophoresis, LD_2 being the most marked.

LD_1 has a higher concentration in tissues with predominantly aerobic metabolism (e.g. heart muscle, red blood cells).

LD_5 has a higher concentration in tissues which are capable of anaerobic metabolism (e.g. skeletal muscle, liver).

Hydroxybutyrate dehydrogenase (HBD):
LD_1 and LD_2 have greater activity towards the substrate hydroxybutyrate dehydrogenase, so measurement of serum HBD represents measurement of these isoenzymes of LDH.

Clinical application of LDH isoenzyme measurement:
1. delayed investigation of suspected myocardial infarction
   — LD_1 remains elevated after total LDH has returned to normal
2. to distinguish further infarction from liver congestion when a second enzyme increase follows one due to myocardial infarction:
   — LD_1 increase → cardiac
   — LD_5 increase → liver

N.B. In myocardial infarction LD_1 equals or exceeds LD_2. This pattern suggests a cardiac origin for LDH but may also be found in:
— megaloblastic anaemia
— acute haemolysis
— renal infarction

Aspartate aminotransferase (AST)
Catalyses the reaction:

$$\text{L-aspartate} + \text{2-oxoglutarate} \rightleftharpoons \text{L-glutamate} + \text{oxaloacetate}$$

AST occurs in generous amounts in heart, liver, muscle and kidney and to a lesser extent in pancreas, spleen and lung.

Causes of increased serum AST:
1. Liver disease (see page 44)
2. Circulatory collapse (shock)
3. Myocardial infarction (see page 165)
4. Acute pancreatitis

5. Cardiac arrhythmias } hepatic congestion
6. Congestive heart failure } hepatic congestion
7. Muscular dystrophy and sometimes other skeletal muscle diseases
8. Pulmonary infarction

Alanine aminotransferase (ALT)
Catalyses the reaction:

L-alanine + 2-oxoglutarate ⇌ L-glutamate + pyruvate

ALT is found predominantly in the liver but also in kidney and to a lesser extent in heart and muscle, pancreas, spleen and lung.

Causes of increased serum ALT:
1. Liver disease (see page 44)
2. Circulatory collapse (shock)
3. Liver congestion secondary to cardiac failure

γ-Glutamyl transferase (γGT)
Catalyses the transfer of γ-glutamyl groups from a natural or synthetic peptide or peptide-like substrate to an acceptor which is usually a dipeptide.

The richest source of the enzyme is kidney with appreciable amounts in liver, pancreas and prostate.

Causes of increased serum levels of γGT:
1. Liver disease (see page 45)
2. Induction of hepatic microsomal enzymes by:
   (i) alcohol
   (ii) some drugs
   (e.g. phenytoin, barbiturates, rifampicin)

5′–Nucleotidase (5NT)
Specifically catalyses the hydrolysis of nucleotides in which phosphate is attached to the pentose ring in the 5-position.

Found in liver, thyroid, aorta and bone.

5NT is useful in the presence of an increased alk. phos. to differentiate:

liver disease — elevated 5NT
bone disease — normal 5NT

N.B. 5NT is not affected by enzyme-inducing agents (cf γGT)

Alkaline phosphatase (Alk. Phos.)
Catalyses the hydrolysis of phosphoric acid mono-esters.

Alkaline phosphatases are present in most tissue, the richest sources being:
1. Bone (osteoblasts)
2. Liver (bile canaliculi)

3. Intestine (epithelium of small intestine)
4. Kidney (proximal tubules)
5. Placenta
6. Breasts during lactation

In all sites they appear to be involved in transport processes across cell membranes.

Causes of increased serum total alk. phos.:
1. Liver disease (see page 44)
   — especially biliary obstruction:
   (i) intra-hepatic (N.B. secondary deposits)
   (ii) extra-hepatic
2. Bone disease (see pages 89–96)
   (i) Paget's disease (osteitis deformans)
   (ii) Tumours
      a. primary—osteogenic sarcoma
      b. secondary (especially from primaries in lung, prostate, thyroid, kidney & breast)
   (iii) Hyperparathyroidism with bone involvement
   (iv) Osteomalacia/rickets
   (v) Renal osteodystrophy
3. Late pregnancy (from placenta)
4. Ectopic production by tumours (see page 140)

N.B. Normal alkaline phosphatase levels are higher in children because of bone growth.

Alk. phos. isoenzymes
The 'isoenzymes' of alk. phos. are not derived from separate genes. Therefore, they do not conform to the strict definition of isoenzymes (see page 158), but they do catalyse the same reaction and it is general practice to call them isoenzymes. It seems likely that their different electrophoretic mobilities are largely related to differences in their sialic acid contents.

The normal isoenzymes are:
1. Liver
2. Bone
3. Intestinal
4. Placental (last trimester of pregnancy only)

The alk. phos. produced ectopically by some tumours is often similar to placental alk. phos. (i.e. the so-called 'Regan' isoenzyme)

Isoenzyme distribution in normal serum:

Adults — liver and bone (± intestinal)
Infants and children — bone predominant
Late pregnancy — placental also present

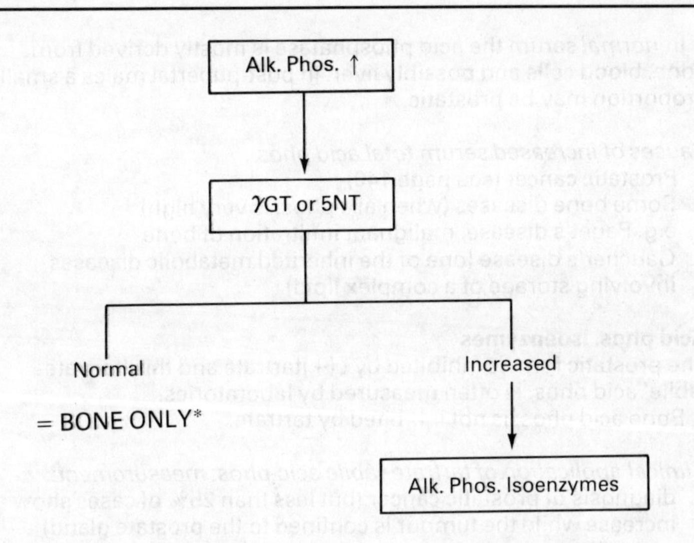

* On rare occasions, a high alk. phos. with normal γGT and 5NT will occur because of:
1. ectopic alk. phos. production by a tumour
or 2. intestinal disease

Fig. 18. Differentiation between the causes of an unexplained elevation of alk. phos.

Clinical application of alk. phos. isoenzyme measurement:
Used to investigate the cause of a high total alk. phos. (see Fig. 18), especially:
1. investigation of high alk. phos. revealed by screening
2. investigation of an apparent elevation where there is already a physiological increase (pregnancy, growing children)
3. differentiation between liver metastases and bone metastases (or both)
4. when γGT is also elevated, and the latter may be related to enzyme induction rather than liver disease

Acid phosphatase (Acid Phos.)
Catalyses the hydrolysis of phosphoric acid monoesters (optimally at an acid pH).

After puberty the prostate is rich in acid phos. Red blood cells have a significant amount and there is some in leucocytes and platelets. Bone (osteoclasts), liver, kidney, spleen and pancreas have minimal amounts.

In *normal* serum the acid phosphatase is mostly derived from bone, blood cells and possibly liver. In post-pubertal males a small proportion may be prostatic.

Causes of increased serum total acid phos.:
1. Prostatic cancer (see page 140)
2. Some bone diseases (when alk. phos. is very high) e.g. Paget's disease, malignant infiltration of bone
3. Gaucher's disease (one of the inherited metabolic diseases involving storage of a complex lipid)

Acid phos. isoenzymes
The prostatic form is inhibited by L(+)tartrate and this 'tartrate-labile' acid phos. is often measured by laboratories.
Bone acid phos. is not inhibited by tartrate.

Clinical application of tartrate-labile acid phos. measurement:
1. diagnosis of prostatic cancer (but less than 25% of cases show an increase while the tumour is confined to the prostate gland)
2. early diagnosis of metastases from prostatic cancer

N.B. Acid phos. may not be produced by very poorly differentiated prostatic tumours even when widely metastasised.

Amylase
Catalyses the hydrolysis of the α -1,4 glycoside bonds of starch leading to eventual production of maltose and maltotriose.
Amylase occurs in the salivary glands and Fallopian tubes as well as in the pancreas.

Causes of increased serum amylase:
1. Acute pancreatitis
2. Pancreatic pseudocyst
3. Morphine administration
4. Mumps
5. Tubal pregnancy
6. Diabetic ketoacidosis
7. Most upper abdominal inflammations e.g. perforated peptic ulcer, cholecystitis } mild increase
8. Renal failure (failure of amylase excretion)

Cholinesterases
True cholinesterase (acetyl cholinesterase)
— found in nerve tissue and erythrocytes.
Pseudocholinesterase (cholinesterase)
— found in liver, heart muscle, intestine and *plasma*

Both catalyse the hydrolysis of a variety of choline and non-choline esters.

Causes of decreased plasma cholinesterase:

1. Inherited abnormal variants
   — there are ten possible genotypes
2. Diseases
   (i) liver disease (liver is the main site of synthesis)
   (ii) pregnancy
   (iii) organophosphate poisoning
   (iv) severe anaemia
   (v) chronic debilitating disease
   (vi) myocardial infarction

Sensitivity to suxamethonium

The muscle relaxant suxamethonium (Scoline) is normally inactivated by cholinesterase in 2–4 mins. Prolonged apnoea after anaesthesia involving the use of this drug may occur in those with reduced cholinesterase activity.

In the case of the inherited variants, the enzymes produced by the abnormal genes show atypical responses to the presence of certain inhibitors (e.g. dibucaine and fluoride). The percentage inhibition by the inhibitor is reported as the *dibucaine and fluoride numbers*. This feature may be used to differentiate the various genotypes.

Genetically normal individuals whose low cholinesterase activity is due to disease show normal dibucaine and fluoride numbers.

ENZYMES IN DISEASE

Myocardial infarction

Following infarction CK, AST and LDH are released into the circulation, peak at different times and are removed at different rates (Fig. 19). (HBD may be measured instead of LDH—see page 160)

Serial measurements of these enzymes are therefore of value but the following limitations should be noted:

1. lack of specificity:
   — CK is also elevated in muscle disease
   — AST rises in liver and muscle disease
   — LDH levels are high in many diseases
2. other chest pains mimicking myocardial infarction:
   — CK }
   — LDH } slightly elevated after pulmonary embolism
3. pre-existing raised enzyme levels due to prevailing disease of other organs
4. secondary rise in enzyme levels due to secondary damage to other organs (e.g. liver following shock or congestive cardiac failure)
5. intra-muscular injections and surgical procedures
   — increase in level of CK

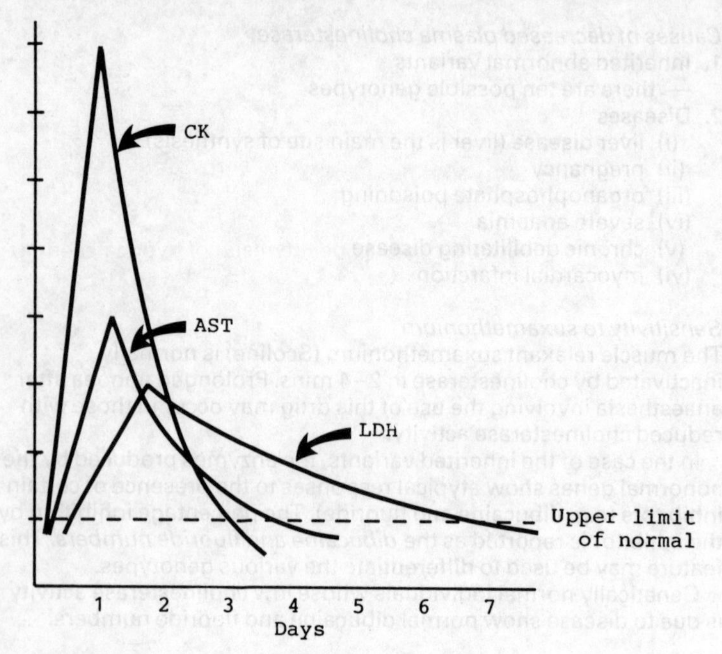

Fig. 19. CK, AST and LDH levels following infarction

6. artefacts
   — haemolysis of sample increases measured level of LDH (or HBD)

Muscle disease
CK shows the greatest elevation in all types of myopathy and is therefore the enzyme of choice for investigation of muscle disease.

Duchenne muscular dystrophy (sex-linked)
— severe and mild forms
— CK levels up to 100 × upper limit of normal (ULN) in the severe form, 10 × ULN in mild form (but wide overlap between the two forms)
— though predominantly CKMM, CKMB can be detected in 90% of cases
— AST & LDH levels usually increased in the severe form
— main uses of CK estimation are:
  1. detection of carriers (female)
  2. diagnosis in 'at risk' siblings prior to clinical presentation

Other muscular dystrophies
Limb-girdle dystrophy (probably autosomal recessive)
— 70% have elevated CK (10–20 × ULN)
— CK estimation fails to detect carriers

Fascioscapulohumeral dystrophy (autosomal dominant)
— CK only slightly or moderately elevated

Polymyositis
— CK elevated in about 70% of patients and may be up to 100 × ULN in children
— steroids usually lower CK (but not in muscular dystrophy)

Possible causes of confusion
CK levels may be elevated, along with AST levels, with muscle damage (e.g. surgery, prolonged epileptic seizures)

Liver disease
The use of enzyme measurement in liver disease is covered in chapter V.

Bone disease
(see pages 89–96)
See also page 162 for the bone diseases which are associated with elevated alk. phos.
In bone disease the level of alk. phos. correlates with osteo*blastic* activity. With osteo*lytic* malignant bone lesions (e.g. multiple myeloma) the alk. phos. is usually normal or only slightly elevated.
Acid phos. may be elevated with some malignant diseases affecting bone (see page 164)

Haematological disease
1. Anaemia
   (i) microcytic (iron deficient)
       — normal enzymes
   (ii) haemolytic
       — slight increase in LDH (LD_1 and LD_2 isoenzymes)
   (iii) megaloblastic
       — marked increase in LDH (up to 20 × ULN in pernicious anaemia)
       — normal or slightly raised transaminases
2. Leukaemia
   — slight increase in LDH (LD_2 and LD_3 isoenzymes)

Therapeutic monitoring and toxicology

THERAPEUTIC MONITORING

Therapeutic monitoring may be required for treatment with the following drugs:

digoxin	procainamide
disopyramide	lignocaine
lithium	quinidine
propranolol	tricyclic antidepressants
theophylline	anticonvulsants
antibiotics	methotrexate

Indications for monitoring

1. To assess compliance
   — non-compliance may be suggested by:
   (i) unexpectedly low or undetectable level
   (ii) conflicting results
   e.g. if a patient on thyroxine defaults and then takes extra drug on the day before clinic, the thyroxine level will be high but TSH will also be high.
2. if the margin between therapeutic and toxic level is narrow or there is a narrow therapeutic range, e.g.

lithium	theophylline*
phenytoin*	methotrexate
digoxin	lignocaine
gentamicin	high dose salicylate*

   * the metabolism of phenytoin, salicylate and possibly theophylline may be saturable within the therapeutic range
3. if clinical findings suggest toxicity
4. if there are no known clinical symptoms of toxicity
5. if the symptoms of toxicity and under-treatment are the same e.g. the antiarrhythmics: quinidine, procainamide
6. in the elderly who may have decreased clearance and/or altered sensitivity
   e.g. digoxin (clearance ↓ sensitivity ↑)
7. in childhood where susceptibility may be increased
   e.g. salicylates

8. in pregnancy where gastric emptying is delayed and absorption time increased
9. in the presence of hepatic, renal or gastrointestinal disease (decreased clearance or malabsorption)
   e.g. methotrexate, gentamicin (renal)
   phenytoin, propranolol (hepatic)
   digoxin (malabsorption)
10. where there is known to be a wide variation in the rate of metabolism of the drug (esp. in children and in old age)
    e.g. propranolol
    theophylline
    tricyclic antidepressants
11. in multiple drug therapy with possible interaction
    e.g.
    (i) the elimination of theophylline is slowed by:
    erythromycin
    cimetidine
    (ii) the half life of methotrexate is lengthened by probenecid

Biochemical assessment of the consequences of drug therapy
Drugs may have unwanted side-effects and biochemical checks are possible to monitor some of these:

(i) long term lithium → hypothyroidism
— check thyroid function tests

(ii) some diuretics, purgative abuse → hypokalaemia*
— check plasma potassium

(iii) hepatotoxic drugs → liver damage
— check transaminases

(iv) nephrotoxic drugs → renal damage
— check plasma creatinine/urea
(or check for proteinuria if the drug produces tubular damage)

* *N.B.* low plasma potassium (or high serum calcium) may increase the sensitivity to digoxin.

Timing of samples for drug assays (Fig. 20)
After each dose the plasma level of a drug will reach a peak and then decline. The time taken for the serum level to halve is called the 'half-life'. Repeated dosage increases the level until an equilibrium is reached between intake and elimination—the 'steady state'. Within this state, peaks and troughs will still occur and the ideal is to have these variations within the therapeutic range.

N.B. it takes approximately 3 to 5 half-lives to reach steady state

Samples for the majority of drugs are taken just prior to the next dose being given (trough level). Some drugs have a very long

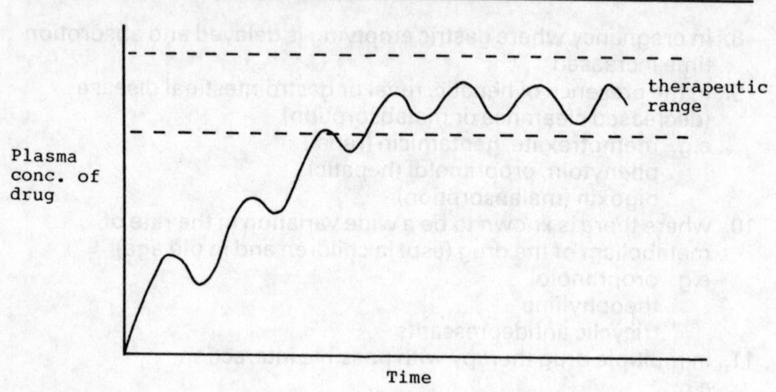

Fig. 20

half-life (e.g. phenytoin) so that timing is not critical, whereas others have a short half-life or are very toxic and then peak and trough levels may be measured.

Digoxin	6–10 hrs. after last dose (best correlation of serum and tissue concentrations)
Valproate	2–4 hrs. after last dose (only really useful to check compliance)
Gentamicin	before and 30–60 mins. after
Theophylline	before and 2–6 hrs. after
Methotrexate	24 hrs. after high dose infusion

Interpretation

Remember that concentrations are meaningless in isolation and for interpretation it is necessary to know:

1. when treatment was started (3–5 × half-life to steady state), e.g.
   — digoxin only reaches steady state 7 days after treatment commenced
   — thyroxine takes about 1 month! (serum half-life is about 7 days)
2. dose regime—what dose and how often?
3. time of last dose
4. time sample was taken
5. age and weight of patient
6. the clinical situation which prompted the request

Action

If plasma level *high* and there is evidence of toxicity—*reduce dose*
If plasma level *low*—check for:
— non-compliance
— poor absorption
— accelerated elimination
and *adjust dose* accordingly.

TOXICOLOGY

Unconsciousness in a patient aged 15–55 with no head injury is most likely to be due to drug overdose.

1. Deliberate overdose
   — frequently:
   benzodiazepines, antidepressants, paracetamol, aspirin } often combined with alcohol
2. Accidental poisoning
   e.g. paraquat, organo-phosphorus pesticides } adults
   iron, vitamins, oral contraceptives } children
3. Iatrogenic
   (i) long term therapy
   e.g. phenobarbitone
   phenytoin
   lithium
   (ii) in the elderly
   e.g. confusion over instructions for medication
   drug interactions (multiple medication)
   increased sensitivity or reduced clearance

Indications for hospital admission in suspected acute poisoning

1. Coma
2. Respiratory depression
3. Shock
4. Hypothermia
5. Convulsions
6. Cardiac arrythmias
7. Known (or suspected) ingestion of drugs with delayed action
   (i) paracetamol (liver toxicity)
   (ii) tricyclic antidepressants (→ cardiac arrhythmias)
   (iii) slow release preparations
   (iv) glutethimide (late apnoea due to accumulation of toxic metabolites)

Investigations

General

1. Blood gases
   (i) to assess respiratory depression
   (ii) to detect an acid-base disturbance induced by the poison
   N.B. cardiac arrhythmias may respond to correction of hypoxia or acidosis

2. Glucose
   (i) to exclude hypo/hyperglycaemia as a cause of coma
   (ii) some poisons may produce hypoglycaemia
   e.g. ethanol
   paracetamol (as a result of liver failure)
3. Electrolytes/urea
   — to assess the effects of vomiting, dehydration etc.
4. Calcium
   — convulsions may be the result of hypocalcaemia

Identification of the poison

Drugs may be detected in:

Blood
Urine
Gastric aspirate
} *Collect all 3 where possible*

Most routine laboratories can detect:

salicylate	— blood
paracetamol	— blood
iron	— blood or gastric aspirate*
lithium	— blood (do not use *lithium*-heparin tube)
ethanol	— blood
digoxin	— blood
paraquat	— urine* or gastric aspirate*

* screening tests only

Confirmation of poisoning by other agents is seldom essential in management as most patients recover with supportive therapy (intensive where necessary) and in less than 1% of cases is there an appropriate 'antidote' (see Table 10).

Attempts to eliminate the poison actively may be indicated in less than 5% of cases (see below), but then qualitative, and where possible quantitative, analysis of the poison is required.

Management

Where an appropriate antidote is available, it is usually important to identify the poison as rapidly as possible. However, when administration of a specific antidote is dependent on the plasma level of the drug, it may be necessary to wait until a few hours after ingestion before doing quantitative analysis (e.g. paracetamol levels peak about 4 hours after ingestion).

Supportive therapy

General supportive therapy (e.g. maintenance of fluid and electrolyte balance, maintenance of respiration etc) is the most important aspect of management in most cases.

Table 10. Antidotes for some poisons

Poison	'Antidote'
Organophosphorus pesticides	Pralidoxime mesylate & atropine
Cyanide	Dicobalt edetate (Kelocyanor)
Iron	Desferrioxamine mesylate (Desferal)
Morphine & analogues e.g. contained in Distalgesic, Lomotil, Fortral, DF118	Naloxone
Paracetamol	N-acetylcysteine (up to 10 hrs from ingestion)
Ethylene glycol (antifreeze) Methanol	Ethanol
Paraquat Diquat	Fuller's earth

Active elimination

1. Emesis and gastric lavage
2. Alkaline diuresis
   (i) may be used for:
   — phenobarbitone and barbitone
   — salicylates
   The rate of tubular reabsorption is decreased in an alkaline environment (when these drugs are maximally ionised) and so excretion is increased.
   (ii) possible biochemical complications:
   a. metabolic alkalosis
   b. hypokalaemia
   c. hypocalcaemia
   d. hypomagnesaemia
   e. hypernatraemia
   (iii) monitor:
   a. plasma drug level
   b. urine pH (to check that adequate alkalinisation is being achieved)
   c. urine electrolytes (to assess losses)
   d. plasma electrolytes
   e. blood gases (acid-base status)
3. Peritoneal dialysis
   (i) less effective than alkaline diuresis for salicylates, phenobarbitone and barbitone but may be used if there is renal impairment

(ii) may also be used for:
lithium
isopropanol
ethylene glycol
methanol/ethanol

4. Haemodialysis
   (i) more efficient than peritoneal dialysis for eliminating:
   lithium
   isopropanol
   ethylene glycol
   methanol
   (ii) effective (but more complicated than alkaline diuresis) for:
   salicylates
   phenobarbitone and barbitone
   (iii) *not good* for:
   short-and medium-acting barbiturates
5. Haemoperfusion
   (i) very effective in eliminating:
   all barbiturates
   some non-barbiturate hypnotics
   disopyramide
   theophylline
   (ii) may help with:
   late presenting paracetamol overdoses
   (iii) possible biochemical complications:
   fall in
   — urea
   — creatinine
   — urate
   — calcium
   — glucose

Appendix

WARNINGS

1. This appendix contains *outlines* of various dynamic function tests. *They are not intended to be complete protocols* for the performance of these tests. For full details, consult the laboratory to which samples will be sent.
2. Those tests involving parenteral administration of compounds may occasionally be associated with anaphylactic reactions and though such reactions are rare, appropriate precautions should be taken
3. **The doses quoted are for adults** unless given on a body surface area or body weight basis

GLUCOSE TOLERANCE TEST

Both the British Diabetic Association and the World Health Organisation now recommend using a 75g oral glucose load.

1. The patient should *not* have been on a low carbohydrate diet for the three days prior to the test
2. The patient should fast overnight. On the day of the test he/she should eat nothing and drink nothing but water.
3. Take a fasting blood sample for glucose estimation
4. Give 75g of glucose dissolved in water. This may be flavoured with squash if desired, or special sachets of flavoured glucose can be used (this makes it more palatable)
5. Take further blood samples for glucose estimation at 30, 60, 90 and 120 mins.

Urine samples:
If it is desired to assess the renal threshold for glucose, urine samples may be collected for glucose estimation basally and at 60 and 120 mins.

SHORT SYNACTHEN TEST

1. This test should ideally be started between 9–10 am as the reference intervals for the response are most well defined for that time. However, useful information can often be obtained later in the day, so it is not ruled out in patients attending in the afternoon
2. Take a basal blood sample for cortisol estimation
3. Give 250 micrograms of tetracosactrin (Synacthen) intramuscularly
4. Take further blood samples at 30 and 60 mins.
   — some protocols only ask for the 30 min sample, but if the response is equivocal at that time, the 60 min cortisol will often show a much higher cortisol level and so exclude deficiency.
   — if sampling is difficult, an alternative is to take a single sample at 45 mins

CLONIDINE STIMULATION TEST

1. Take a basal blood sample for growth hormone
2. Give an oral dose of clonidine (0.15 mg/m^2 body surface area)
3. Take further blood samples for growth hormone at 30, 60, 90, 120 and 150 mins

N.B. clonidine may cause hypotensive effects in some patients

TRH TEST

Contraindications:
chronic obstructive airways disease, asthma, ischaemic heart disease, pregnancy

1. Take a basal blood sample for TSH estimation
2. Give 200 micrograms of TRH intravenously
3. Take further blood samples for TSH at 20 and 60 mins

In most cases of suspected *primary* thyroid disorder, the 60 min sample provides little extra information and can be omitted.

LHRH TEST

1. Take a basal blood sample for LH estimation
2. Give 100 micrograms of LHRH intravenously
3. Take further blood samples for LH at 20 and 60 mins

INSULIN STRESS TEST

Contraindications:
ischaemic heart disease, epilepsy, severe unequivocal panhypopituitarism

Precautions:
(i) the patient must be under *constant medical surveillance*
(ii) 50% glucose must be available for i.v. use should the patient develop profound or prolonged hypoglycaemia

1. Fast the patient from midnight (water allowed)
2. Take a basal blood sample for glucose and the hormones being tested i.e. growth hormone and/or cortisol (this test can be used to assess both growth hormone and ACTH release from the pituitary; ACTH release is assessed indirectly by measurement of cortisol response)
3. Give insulin intravenously. The dose depends on the clinical situation:

suspected hypopituitarism	0.1 units/kg body weight
probable normals	0.15 units/kg body weight
suspected insulin resistance*	0.3 units/kg body weight

   * e.g. obesity, acromegaly, Cushing's syndrome
4. Record the time and nature of hypoglycaemic symptoms. The patient should sweat when hypoglycaemic, but this usually only lasts 10–15 mins. If the patient develops angina, becomes shocked or cannot answer simple questions I.V. glucose *must* be given, but continue sampling *from another site*
5. Take blood samples at 30, 45, 60 and 90 mins for glucose and the hormone(s) being tested
6. After the test, a glucose drink and food should be given. Ensure adequate meals for the rest of the day

TRIPLE PITUITARY STIMULATION TEST

The insulin stress test, TRH test and LHRH test can be combined into a single test. Contraindications and precautions are as given for the individual tests (see above). The TRH and LHRH can be given simultaneously (Relefact contains the appropriate doses in a single ampoule). The insulin should not be mixed with these, but can be given immediately before or after. Subsequent venous samples are then taken at the appropriate times for the individual tests.

WATER DEPRIVATION TEST

If the osmolality of an early morning urine is > 600 mmol/kg this test is not necessary.

There are many variants of this test, but the following is typical:
1. No previous overnight fluid restriction
2. Prior to the commencement of the test the patient may have a light breakfast but no tea or coffee
3. Smoking and alcohol consumption are prohibited.

4. The patient should be weighed just before starting the test and then every time urine is passed. If there is loss of more than 3% of the initial body weight, the test should be stopped and blood and urine sent for osmolality measurement at that time.
5. No fluids are allowed once the test period has commenced, but dry food is allowed.
6. At 0800 hours, urine is passed and this specimen discarded
7. Collect samples and perform measurements as shown below:

URINE			BLOOD	
Time	Measure volume	Measure osmolality	Time	Measure osmolality
0800–0900	+	+	0830	+
0900–1000	+			
1000–1100	+			
1100–1200	+	+	1130	+
1200–1300	+			
1300–1400	+			
1400–1500	+	+	1430	+
1500–1600	+	+	1530	+

8. After 1600 hrs the patient may drink

DESMOPRESSIN TEST

This test should not be performed on patients with primary polydipsia (e.g. compulsive water drinking) because of the danger of water intoxication. Hence it is usually preferable to await the results of the water deprivation test before proceeding.

1. During the test the patient may be allowed to drink fluids as required to satisfy thirst, but *fluids must not be forced* because of the danger of water overload
2. The patient should empty his/her bladder and discard this urine
3. Collect a basal urine for osmolality

   } not necessary if this test follows immediately after a water deprivation test (items 2 and 3)
4. Give 2 micrograms of desmopressin (DDAVP) intramuscularly
5. Collect four consecutive one hour urine collections, measure the volume of each and send an aliquot for osmolality

URINE ACIDIFICATION TEST

Precautions:
This test should not be performed on patients who are already acidotic (there is no point and it can be dangerous)

1. Give ammonium chloride 0.1 g/kg body weight as gelatine capsules or a mixture. It should be given 0.5 or 1.0 g at a time,

during a meal, over a half hour period. The dose is usually taken during breakfast
2. Collect urine at 2, 4 and 6 hours and send each to the laboratory immediately after collection, for pH measurement
3. Collect a capillary blood sample at 4 hours to ensure that an acidosis has been achieved

BROMSULPHTHALEIN (BSP) EXCRETION TEST

This test is not indicated if routine liver function tests other than bilirubin are abnormal.

1. Fast the patient overnight
2. Inject 5 mg/kg body weight of BSP intravenously during a 30 second interval, *taking great care to avoid extravasation* (BSP is extremely irritant)
3. Take a blood sample for BSP estimation from the opposite arm 45 mins later.
4. If Dubin–Johnson syndrome is suspected, take a further sample at 90 mins.

N.B. the presence of ascites, congestive cardiac failure or albuminuria may affect BSP retention.

Index